PREFACE

Basic Transistors is the product of several years of research and experimentation with new teaching methods at the New York Institute of Technology. Because of the increasing importance of the transistor as a substitute for the vacuum tube a good deal of material related to the use of transistors in electronic applications has been introduced.

As with the techniques of *Basic Television*, *Basic Transistors* is highly pictorialized and presents only one important concept per page. Question and Problem pages appear through the volume to emphasize the major points presented.

This simplified, heavily pinpointed approach makes *Basic Transistors* readily understandable without an instructor, making it suitable for individual as well as classroom study. Coverage ranges from the structure of atoms to the characteristics and circuitry of recently developed transistor types.

I should like to acknowledge the assistance of the staff of the New York Institute of Technology, and in particular Mr. Harvey Pollack, who organized the text material.

ALEXANDER SCHURE

New York, N. Y.
February 1961

PREFACE

basic
transistors

by Alexander Schure, Ph.D., Ed.D.

Author of *Basic Television*
Editor of *Electronic Technology Series*

JOHN F. RIDER PUBLISHER, INC., NEW YORK

Elroy Brace

CONTENTS

Preface iii

Atoms and Semiconductors 1

P-N Junctions 18

Questions and Problems 32

Biasing the Transistor 33

Transistor Characteristics 39

Circuit Configurations 57

Questions and Problems 61

Transistor Amplifiers —— Common-Base 62

Transistor Amplifiers —— Common-Emitter 72

Transistor Amplifiers —— Common-Collector 87

Sample Circuits 94

Questions and Problems 98

Coupling Transistor Circuits 99

Power Transistors 118

Questions and Problems 127

Transistor Oscillators 128

Tetrode Transistor 137

Questions and Problems 142

Index 143

What a Transistor Is

The transistor is an amplifier of electric currents. It differs from the vacuum tube (also an amplifier) in that it is made of matter in the solid state. Amplification in a vacuum tube occurs as a result of our ability to control the movement of *electrons* in the empty spaces within the glass envelope. By making proper use of such a controlled electron flow, we can build circuits in which the variations of output current, are greater than the variations of input current. This, of course, is amplification.

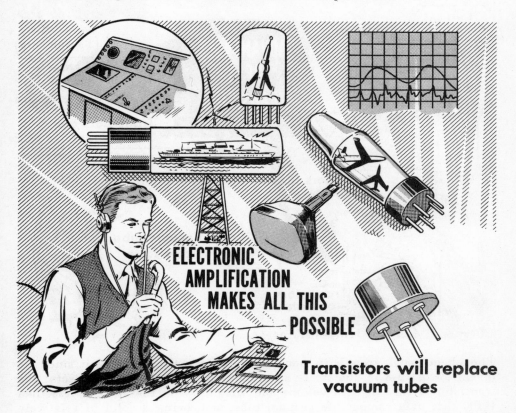

ELECTRONIC AMPLIFICATION MAKES ALL THIS POSSIBLE

Transistors will replace vacuum tubes

Forty-two years after Dr. Lee DeForest invented the amplifying triode tube, scientists at the Bell Telephone Laboratories discovered how to obtain similar effects in certain *solid* materials. This discovery marked the first significant departure from established vacuum tube practice since DeForest's work. Officially credited with the invention of the transistor, W. Shockley, W. H. Brattain, and J. Bardeen of Bell Labs showed that it was possible to construct a solid state device that could perform any function of a vacuum tube. Furthermore, this device — the transistor — immediately showed great promise of having very important advantages over the vacuum tube.

Advantages of Transistors

Vacuum tubes are relatively large in size, mechanically fragile, produce a considerable amount of heat that represents wasted power, and require relatively frequent replacement in normal use.

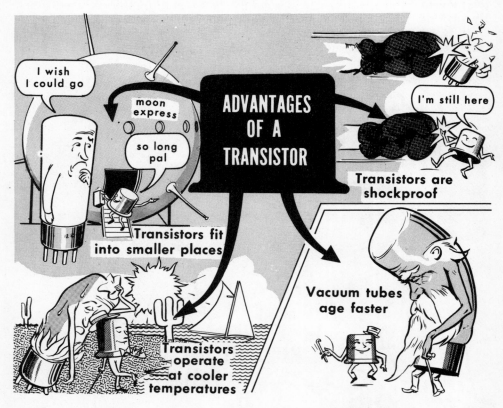

The phenomenal growth in popularity the transistor has seen since its invention in 1948 is due to its advantages in the following respects: the transistor is tiny in both volume and weight — on the average less than 1/100 that of a comparable vacuum tube, and, being a solid of relatively small mass, the transistor can undergo severe mechanical shock without noticeable effect. Where many vacuum tubes use more electrical power merely to heat their filaments than they produce in the form of useful output, transistors require no heater current at all. This one fact alone permits operating efficiencies that were thought impossible years ago. The portable battery-operated radios of today, using five, six, or even seven transistors instead of tubes, can run many hours on the power provided by flashlight cells. Finally, with reasonably careful treatment, a transistor has an indefinite life since it contains nothing that can wear out.

Understanding Transistors

The invention of the transistor could not have come much earlier in the history of the human race than it did. Deep forages into the mysteries of nature cannot be made without ample knowledge of nature's fundamental structure and basic physical laws. These great conquests of science are not accidental. As we learn more of the character of our universe's foundation, we can ascend ever higher into its superstructure with confidence and certainty.

Thus, to understand the functioning of transistors, we must first investigate how matter is made, how it behaves under normal and natural conditions, and what we can do to it to make it do our bidding. No one can hope to reach the stars merely by stretching his arms outward into space toward them. To appreciate and apply transistors, we must first learn something about the structure and behavior of *matter* in its natural state.

Someday this man will reach the stars through basic research

This man will never reach the stars

The Universe of Matter

Matter is the stuff of which the universe is made. As we look around us, we can count literally thousands of different material items: the leaf of a

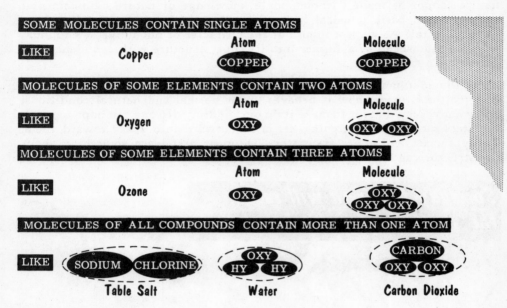

tree, a blade of grass, a grain of sand, the knob on a door, a bird on wing, and so on. Yet for many years we have known that all objects can be reduced to combinations of basic chemical *elements,* many of which are familiar to us. Iron, carbon, oxygen, and gold are elements. When chemically combined they form *compounds* — like carbon dioxide, water, and iron rust. From a little over 90 elements, nature has formed thousands of compounds.

Consider a lump of the element, silver. If we had a microscope powerful enough, we would find the lump consists of extremely small units, each of which possesses all the characteristics of silver. These units are called *molecules.* But if we examine a compound like water, we discover that each molecule of water contains three smaller units of matter: two of these are hydrogen, the other is oxygen. Such molecular building blocks are called atoms, so that a molecule of water is composed of two atoms of hydrogen and one of oxygen. For many *elements* in which one molecule contains a single atom, the molecule and atom are identical. This is true of metals, in particular. Other elements, however, are built of *diatomic* molecules. This means that each molecule consists of two identical atoms, as in the case of chlorine, hydrogen, oxygen, nitrogen, and many other elemental gases. Thus, the molecule of an element may contain one or more atoms while the molecule of a compound always contains two or more atoms.

Atomic Particles

We have spoken of atoms as units used by nature to build the matter of the universe. As we examine them more closely — not by actually seeing them, for this we cannot do — we find that even these tiny bodies are made of still smaller parts. This information comes to us from observations of their behavior in large groups. By applying certain immutable physical laws to these observations, we are forced to conclude that all stable atoms (except hydrogen) contain three basic particles: *electrons, protons,* and *neutrons.* Hydrogen is different in structure from all other elements in that its atom is composed of only two particles — one electron and one proton. Let us look inside a hydrogen atom to study its structure.

Although modern science no longer conceives of an atom as a miniature solar system with planets (electrons) whirling around a central sun (protons and neutrons in the nucleus), this point of view is helpful in visualizing atomic structure for study at a basic level. The hydrogen atom is the

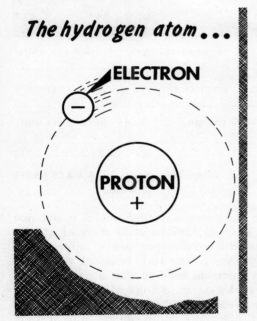

The hydrogen atom...

ELECTRON

PROTON
+

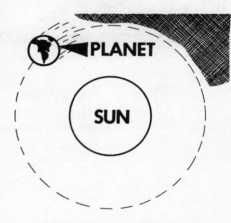

PLANET

SUN

... is like a solar system containing a single planet revolving around a central sun

simplest system of all. At its center is a single proton, a particle with one positive electrical charge. Revolving around it, in a definite fixed orbit, is its *negative* counterpart — a single electron. Viewed from the outside, a hydrogen atom appears to be electrically neutral because the + and − charge values are identical in size and therefore cancel each other, leaving the net atomic charge equal to zero.

More Complex Atoms

Next in line is the helium atom. This element is built of atoms that contain two planetary electrons, both revolving in the same orbit around a nucleus at the center. Since a helium atom is electrically neutral (as are the atoms

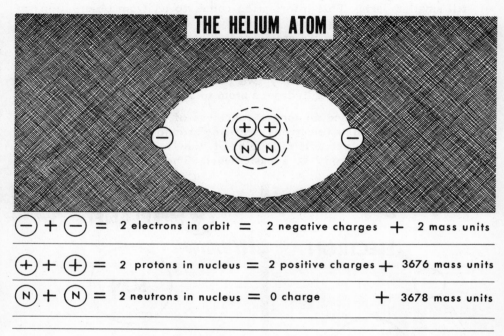

THE HELIUM ATOM

$\ominus$ + $\ominus$ = 2 electrons in orbit = 2 negative charges + 2 mass units

$\oplus$ + $\oplus$ = 2 protons in nucleus = 2 positive charges + 3676 mass units

N + N = 2 neutrons in nucleus = 0 charge + 3678 mass units

TOTALS   0 CHARGE   AND 7356 MASS UNITS

of *all* elements), there must be at least two protons in the nucleus to balance the negative charge of the external electrons. In the early days of atomic investigation, it was found by several different experimental approaches, that electrons contribute scarcely anything to the total mass of the atom. If we assign a mass of one unit to an electron, we find that a proton has a mass of 1838 of the same units. Careful determinations of the mass of a helium atom disclosed, however, that it contained 7352 such units, or *four* times the weight of a single proton. Since there could not be four protons in the nucleus with only two electrons in orbit (because this would form a positively charged atom), it was concluded that there must be another kind of atomic particle having the same mass as a proton but without any kind of charge. This was verified in 1930 with the discovery of the *neutron*. By placing two neutrons together with the two protons in the helium nucleus, all the facts obtained by charge and mass measurements could be explained.

More Complex Atoms

The essential simplicity of nature is demonstrated by the fact that as we add more and more complex atoms to our list, we find that a new element is formed every time *one* additional electron is added to orbiting ones already there, with *one* more proton placed in the nucleus to maintain electric neutrality. The same easy sequence is not true of neutrons, however, because the increase of relative atomic mass does not follow uniformly from one atom to the next.

Hydrogen, the simplest element, is identified in the order of elements by assigning to it the number *1*. This is called its *atomic number*. On this basis, helium is atomic number *2*. The relative mass of each atom, measured in terms of comparative units, is called the *atomic weight* of the atom, and is determined by the proton plus the neutron content. Another point of importance is that all the revolving electrons do not move about at the same distance from the nucleus. As we go higher in atomic number, we find that the electrons tend to congregate in different orbits. After helium comes *lithium* with a total of three electrons, two of them in the first ring (or shell) and one in the second. For reasons that are fairly well established, nature forces electrons into shells that give the atom the utmost stability.

THE FIRST ELEVEN ATOMS

Symbol	Element	At. No.	At. Wt.	protons	electrons	neutrons
H	hydrogen	1	1	1	1	0
He	helium	2	4	2	2	2
Li	lithium	3	7	3	3	4
Be	beryllium	4	9	4	4	5
B	boron	5	11	5	5	6
C	carbon	6	12	6	6	6
N	nitrogen	7	14	7	7	7
O	oxygen	8	16	8	8	8
F	fluorine	9	19	9	9	10
Ne	neon	10	20	10	10	10
Na	sodium	11	23	11	11	12

7

Atoms of Electrical Conductors

The forces that hold an atom together as a unit are often called *short-distance* forces because they are very effective over short distances, but become rapidly ineffective as the distances are increased. Thus, in the *sodium* atom, the two electrons in the inner shell are very tightly bound to the nucleus by strong short-distance forces. The single outer electron in sodium, however, is so far from the nucleus that it can be dislodged easily. When atoms are bombarded by energetic radiations such as cosmic rays, free flying protons, electrons, or X-rays, the outermost electrons are the first to be stripped from the atoms. These vulnerable outer electrons are referred to as *valence electrons*. Thus, the valence electron in lithium is the only·one that has started the second shell. In the case of sodium, the valence electron is the eleventh one — this one having begun a *third* outer shell.

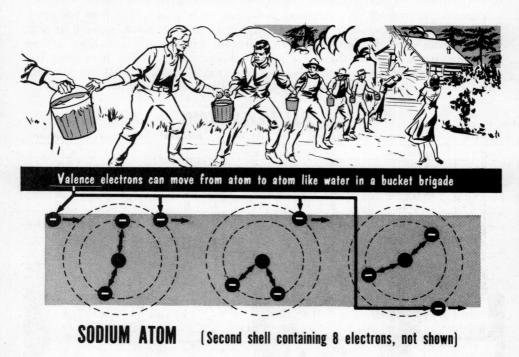

Valence electrons can move from atom to atom like water in a bucket brigade

SODIUM ATOM (Second shell containing 8 electrons, not shown)

Most metals (like silver, copper, aluminum, etc.) release their valence electrons without too much persuasion. When this happens, the body of the metal is filled with *free electrons* that can move from atom to atom like the water in a bucket brigade. Therefore, the free electrons serve as current carriers, making it possible for an electric charge placed at one end of a wire to make its effects felt at the other end. Metals therefore, are *conductors* of electricity.

Atoms of Electrical Insulators

Referring to the table of the first eleven elements, we note that two of these, chemically, are unique. Both helium and neon are members of the family of *inert elements*. This means that they refuse to give up or take on ad-

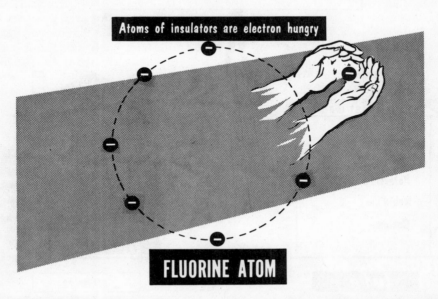

Atoms of insulators are electron hungry

FLUORINE ATOM

ditional electrons in chemical reactions. For this reason, there are no common compounds of either neon or helium — they simply refuse to share electrons (either by giving or taking) with other elements. These observations lead to the inevitable conclusion that certain shell configurations are so stable that nothing we can do, under normal circumstances, will force electrons in or out of these atoms. The first shell is said to be *filled* or stable with two electrons; the second shell is filled when it contains eight electrons. Throughout all of the chemical elements this condition of preferred outer ring content is always in evidence. We have seen that a metal like sodium has one *extra* electron beyond the filled condition. This is another reason we might use to explain its willingness to give up one electron. Since, by doing so, its outer ring reverts back to the stable condition of eight electrons.

An element like fluorine, is shy one electron of the filled-eight count. When we try the bucket brigade with such atoms, they simply grab the electron being passed along, and hold tight. All such elements display great reluctance to part with the electron once it has entered the atom. Elements like fluorine, sulfur, and iodine (those needing an electron or two to complete their outer rings) do not provide electrical current carriers and therefore behave as *insulators*.

The Semiconductors

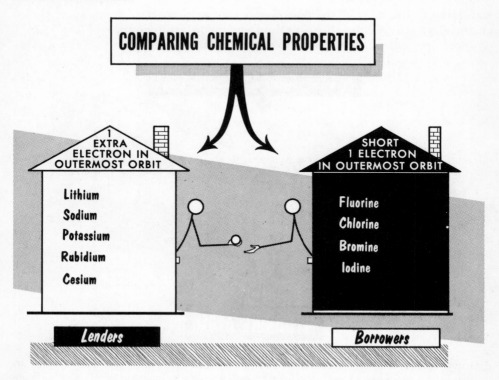

COMPARING CHEMICAL PROPERTIES

1
EXTRA
ELECTRON IN
OUTERMOST ORBIT

Lithium
Sodium
Potassium
Rubidium
Cesium

Lenders

SHORT
1 ELECTRON
IN OUTERMOST ORBIT

Fluorine
Chlorine
Bromine
Iodine

Borrowers

The table of the first eleven elements discloses another interesting relationship. Of the first eleven elements, three have one particularly outstanding feature in common: hydrogen (atomic No. 1), lithium (atomic No. 3) and sodium (atomic No. 11) each has one electron in the outermost orbit. When the chemical properties of these elements are compared, the similarity is very striking. Chemists have no doubt that such elements are related in chemical activity *because of this common feature.* All such elements are therefore placed in *one family.* Similarly, fluorine, chlorine, bromine, and iodine are members of the same family because each of these elements is short one electron to complete a stable outer shell. This family is completely different from the H, Li, Na group because of the opposite nature of its electron configuration.

Midway between these two groups is a family of elements that displays the properties of both groups to varying degrees. Our chart of the 11 elements shows that carbon has 4 electrons in its outer shell. On the basis of previous reasoning, you might suspect that this element would be content if it could either gain 4 electrons to complete its outer shell, or lose 4 to revert back to an inner shell that is complete with 2 electrons.

ATOMS AND SEMICONDUCTORS

The Semiconductors (contd.)

Copper and silver are among the electron lenders. These metals are always good electrical conductors. At the opposite extreme, manganese and iodine — electron borrowers — are always very poor conductors. What of carbon and other members of its family? If we look through a list of elements to find others that have 4 electrons to lend, or that have an outer shell that needs 4 electrons to complete it, we should find that both germanium and silicon fit into these classifications. These are the commonly used transistor materials. What makes them so specially suited for use in transistors?

Let us look more closely at the atomic picture that carbon presents. This element occurs in two natural forms: black graphite and the crystalline

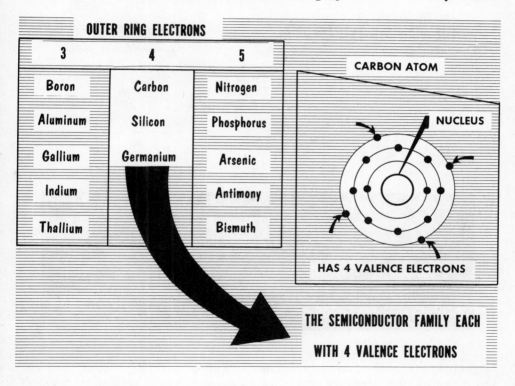

OUTER RING ELECTRONS

3	4	5
Boron	Carbon	Nitrogen
Aluminum	Silicon	Phosphorus
Gallium	Germanium	Arsenic
Indium		Antimony
Thallium		Bismuth

CARBON ATOM

NUCLEUS

HAS 4 VALENCE ELECTRONS

THE SEMICONDUCTOR FAMILY EACH WITH 4 VALENCE ELECTRONS

form we call *diamond*. In its black or noncrystalline state, carbon is an excellent electrical conductor — perhaps not as good as the average metal, but good enough to be used in the manufacture of dry cells, carbon microphones, and projector arcs. From what we have previously said, this must mean that the 4 outer electrons (valence electrons) are not tightly bound to the atomic nucleus and can move about freely as current carriers.

The Semiconductors (contd.)

Should we try to substitute the crystalline form of carbon for the noncrystalline form in an electrical circuit, we should find the carbon has become an *extremely poor conductor*. Evidently, some structural change has occurred that has caused the valence electrons to become fixed in place so that they are no longer free to move.

This can mean only one thing: when carbon atoms join to form a crystal, their valence electrons must come under the influence of some new kind of binding force that does not exist in the noncrystalline form. Evidence obtained from X-ray scattering in crystals, demonstrates that the atoms line up in a geometrically perfect lattice that has extreme stability. A new kind of *bond* has been created. The valence electrons, in the grip of their

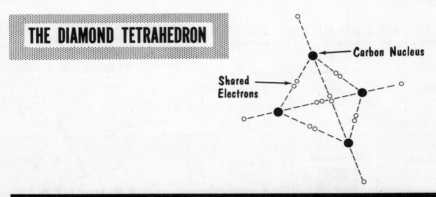

THE DIAMOND TETRAHEDRON

Carbon Nucleus

Shared Electrons

EACH CARBON NUCLEUS SHARES AN ELECTRON WITH EACH OF THREE OF ITS NEIGHBORS IN ITS OWN TETRAHEDRON, LEAVING ONE ELECTRON OF EACH ATOM FREE TO JOIN WITH CARBON ATOMS OF NEIGHBORING TETRAHEDRONS. THUS INDIVIDUAL CRYSTALS MAKE UP A LATTICEWORK OF BONDED CRYSTALS IN WHICH EVERY ELECTRON IS SHARED. THIS MAKES FOR HIGH STABILITY OF STRUCTURE.

neighbors, are now *covalently bonded*. (The prefix "co" means "shared" and we think of a diamond crystal as being made up of carbon atoms, each sharing its valence electrons with other nearby atoms.) From various studies of crystals, it has been learned that the diamond crystal takes the shape of a perfect tetrahedron with a carbon atom in each corner. Then each tetrahedron teams up with neighboring tetrahedrons so that all four of the valence electrons of every carbon atom are shared with other carbon atoms.

The Semiconductors (contd.)

The tetrahedron is a three-dimensional figure and has been so represented in the preceding drawing. When we attempt to picture many such inter-

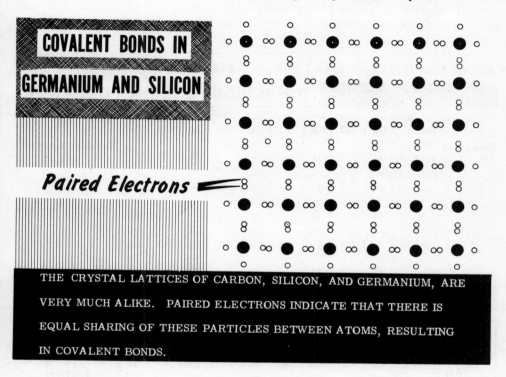

COVALENT BONDS IN GERMANIUM AND SILICON

Paired Electrons

THE CRYSTAL LATTICES OF CARBON, SILICON, AND GERMANIUM, ARE VERY MUCH ALIKE. PAIRED ELECTRONS INDICATE THAT THERE IS EQUAL SHARING OF THESE PARTICLES BETWEEN ATOMS, RESULTING IN COVALENT BONDS.

linked tetrahedrons, we find that the result is complex and confusing. To make the study of semiconductors somewhat easier, it is much more convenient to diagram the tetrahedrons in two, rather than three dimensions, showing only the horizontal bonds between atoms in the plane of the paper. Remember at all times, however, that a crystal bar of usable dimensions is made up of millions of tetrahedrons bonded together upward and downward as well as to the right and left.

Both *germanium* and *silicon*, the other two semiconductor members of the carbon family, form crystal lattices very similar to that of carbon. The lattices are symmetrical, being composed of atoms linked by covalent bonds. Thus, both of these semiconductors exhibit exactly the same electrical properties as diamond when they are in crystalline form: germanium and silicon crystals in the *pure state* are nonconductors when used at reasonably low voltages. In the pure state, therefore, neither of these is suitable for use in making transistors.

Adding Arsenic or Antimony to Germanium

If we were to measure the resistance of a bar of pure germanium about one centimeter long and one square centimeter in cross-sectional area, we would find its resistance to be of the order of hundreds of thousands of ohms. Should we now add a piece of arsenic or antimony (the size of a pinhead)

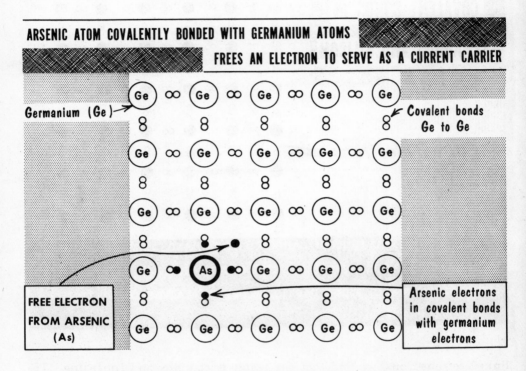

ARSENIC ATOM COVALENTLY BONDED WITH GERMANIUM ATOMS

FREES AN ELECTRON TO SERVE AS A CURRENT CARRIER

Germanium (Ge)

Covalent bonds Ge to Ge

FREE ELECTRON FROM ARSENIC (As)

Arsenic electrons in covalent bonds with germanium electrons

to molten germanium and let it recrystallize, we would find that the same bar would now show a resistance of less than 100 ohms.

Evidently, the added "impurity" does something to destroy the almost invulnerable crystal lattice, thereby making available electrons to act as current carriers. Looking back at the chart showing the carbon family surrounded by other families, we see that both arsenic and antimony belong to the family in Group 5A. This group, unlike the carbon family, has 5 orbital electrons in the outermost atomic shell. What happens when a *pentavalent* atom (containing five valence electrons) such as arsenic or antimony finds itself surrounded by *tetravalent* atoms (containing four valence electrons) of germanium? The arsenic atom also likes to form a covalent bond. As it takes its place in the germanium lattice, however, only four of its five electrons can find bonding partners. This leaves one free electron without a mate.

ATOMS AND SEMICONDUCTORS

The Effect of the Added Impurity

Suppose we connect a source of electric potential across our "impure" bar of germanium. Although the extent of the impurity is perhaps one atom of arsenic to every 12,000,000 germanium atoms, we should find that a current can now flow through the bar with much less opposition than before. That is, the resistance of the germanium has undergone a significant decrease.

Here again, our visualization of the events that take place inside the impure germanium is a matter of educated conjecture. We are pretty sure of the accuracy of the conjecture because it fits into the experimentally observed facts. We believe that the free electron released by the arsenic as it enters into a combination with a germanium atom is attracted toward the positive

THE IMPURITY BUCKET BRIGADE

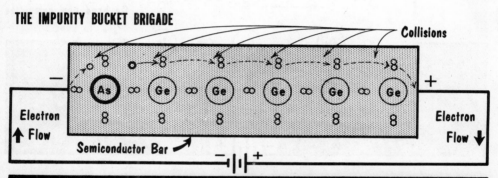

SUCCESSIVE COLLISIONS BETWEEN ELECTRONS CAUSE A DRIFT OF REPLACED ELECTRONS TOWARD THE POSITIVE TERMINAL OF THE BATTERY. AS AN ELECTRON LEAVES THE BAR TO MOVE INTO THE BATTERY, A NEW ONE ENTERS FROM THE OTHER SIDE TO TAKE ITS PLACE. THUS, THE ARSENIC ELECTRONS ARE NEVER USED UP.

terminal of the voltage source by the electric field set up within the bar. The electron cannot move very far, however, before it collides with another electron in a nearby germanium atom. As a result of the collision, the atomic electron is replaced by the moving free electron. What was a free electron before now becomes part of a covalent bond. The newly freed electron then moves to the next atom, ejecting a fresh electron from the atom and taking its place. Thus we again have a "bucket brigade", this time started by a single free electron from an arsenic atom. Once an impurity is added to germanium, it is forever after a true semiconductor with a much lower resistance than the pure material.

Adding Indium to Germanium

A similar reduction of the resistance of pure germanium occurs when we add tiny quantities of elements such as gallium or indium. Referring once again to the listing of elemental families around the carbon group, we note that both gallium and indium are in group 3A. These are *trivalent* atoms having three valence electrons. Specifically, each atom is one electron short of having enough to enter into covalent bonding with germanium.

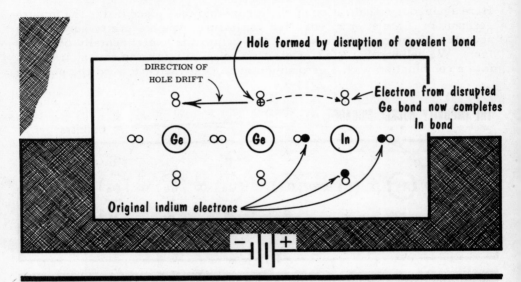

Creation and movement of a "hole" in germanium with an indium impurity

For example, when an indium atom is fed into a pure germanium crystal, the indium replaces a germanium atom from the lattice *only* if it can adopt a fourth electron from a neighboring germanium atom to complete its four covalent bonds. As the fourth electron is snatched from its former position, the position it originally occupied becomes a region of positive charge or a *hole*. The size of the positive charge is about the same as that of an electron, only opposite in sign.

Now suppose that a voltage is applied across the bar. The resulting electric field will cause the newly formed hole to drift toward the negative terminal of the voltage source. Just as there was a "bucket brigade" start in the previous case with electrons, we find that holes exhibit the *same kind of mobility* — except that they move in a direction opposite to that of free electrons.

Electrons and Holes

In discussing the movement of electron current in a conductor such as copper, there is no need for the hole concept because this element contains many free electrons ready to act as current carriers. In a semiconductor containing a trivalent impurity, however, holes take on real meaning. For a better understanding of transistor action, we should think of a hole as a real positive particle, just as an electron is a real negative particle. As the bucket-brigade of holes moves from the positive toward the negative end of a germanium bar, we should think of this motion as an electric current that has the same effect as an electron flow on the external load circuit, except that it is opposite in polarity. Thus, if holes flow from right to left in a given germanium bar, the external load will "see" the same thing as though an electric flow had taken place from left to right. In other words,

AN EXTERNAL INDICATOR CANNOT TELL WHETHER CURRENT CARRIERS ARE HOLES OR ELECTRONS

Germanium + arsenic current carriers are electrons

INDICATOR

SAME INDICATION

Germanium + indium current carriers are holes

INDICATOR

there is no way to place an indicator (meter, oscilloscope, etc.) in the load circuit that will tell us whether holes or electrons in motion in the germanium are causing the current we observe. We might well ask what difference it makes *which* carriers are actually responsible for the current since the effect upon the load is the same in either case. The answer is that we need the *hole concept* to help us explain how transistors amplify — as will be explained further in this book.

P and N Germanium

It is helpful at this time to summarize some of the concepts discussed thus far and to add some new terms to our transistor vocabulary.

The addition of a *pentavalent* impurity such as arsenic or antimony produces a type of transistor germanium in which the charge carriers are *electrons*. Because the impurity adds or donates electrons to the crystal lattice, it is known as a *donor* impurity and the impure germanium itself is designated as *n-type* germanium. The n stands for *negative*, describing the charge characteristics of the carrier — electrons in this case.

A *trivalent* impurity such as indium or gallium is known as an *acceptor* impurity because it must take, or accept, electrons from surrounding germanium atoms in order to enter the covalent bonding scheme. Since the charge carriers in this case are *positive* charges, or holes, transistor germanium to which acceptor impurities have been added is called *p-type* germanium.

Forming a Junction — Diffused-Alloy Process

As we shall see later, transistor action is generally obtained through use of junctions between n-type and p-type germanium sections. Such junctions must be intimate ones in which the n-type and p-type germanium are very closely associated. Several processes are used to secure efficient, high-performance junctions; among the most common of these is the *diffused-alloy process*.

A very thin section of one type — say n-type — of germanium is mounted in a holder and tiny indium dots pressed into each side. The entire assembly is then placed in a carefully controlled furnace where the indium melts and gradually diffuses below the surface of the n-type germanium. A germanium-indium alloy thus forms on each face, converting these to p-type

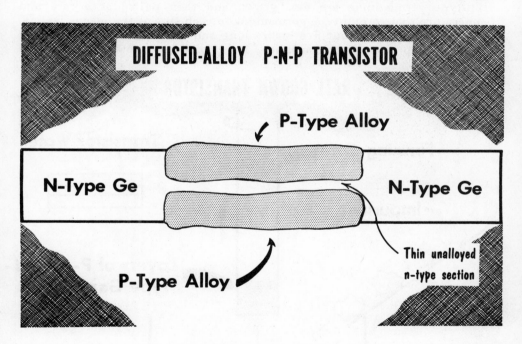

DIFFUSED-ALLOY P-N-P TRANSISTOR

P-Type Alloy

N-Type Ge

N-Type Ge

Thin unalloyed n-type section

P-Type Alloy

germanium. During the heating process, the depth of penetration of the alloy material is made great enough so that the remaining unaffected n-type germanium is an *exceedingly thin* section.

Thus, the diffused-alloy process is often used to produce *p-n-p* transistors. This designation is used to indicate that the "sandwich" contains two junctions (shown by the two hyphens in p-n-p) and that the outer layers are p-type transistor semiconductor while the inner layer is n-type material.

Forming a Junction — Rate Growth Process

The *rate grown transistor junction* is formed by a completely different technique. When a molten mass of a normally crystalline solid is allowed to cool, crystals form automatically. If the cooling is rapid, the crystals will be small; if allowed to cool slowly, the crystals may become quite large.

In the rate growth process, molten germanium is doped with impurities so proportioned that, although the original molten mass is p-type germanium, it changes to n-type germanium when the temperature of the mass is carefully controlled. The conversion from one type to the other is also dependent upon the rate of "pulling" (the rate at which the forming ingot is withdrawn from the "mother" bath as the crystals slowly grow on the ingot). By suitable growing techniques, a dozen or more very thin layers of p-type germaninum are first grown and then pulled at a rate and temperature that form n-type germanium on each side of the p-type wafers. The resulting ingot, about five inches long and as thick as your thumb, is then sawed up into n-p-n assemblies. It is possible to obtain up to 10,000

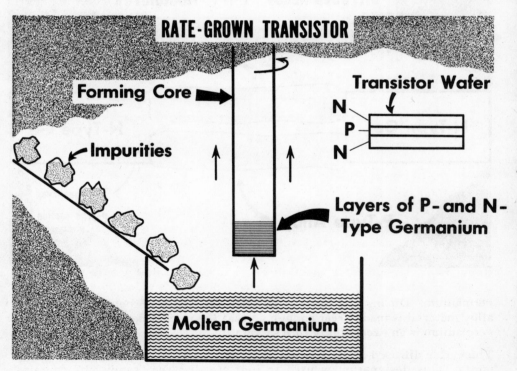

transistor wafers from a single ingot in this way. The resulting "sandwich" in this case contains a very thin p-layer between two much thicker n-layers to form an n-p-n transistor.

Forming a Junction — Gaseous Diffusion Method

The *gaseous diffusion method* of junction formation is confined mostly to the manufacture of large power transistors commonly found in the output stage of automobile receivers and in industrial power amplification systems.

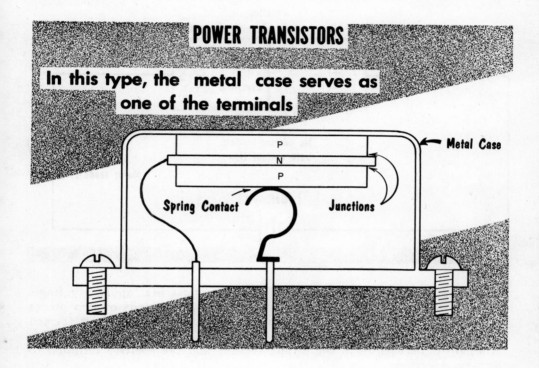

POWER TRANSISTORS

In this type, the metal case serves as one of the terminals

Metal Case

P
N
P

Spring Contact   Junctions

To form a gaseous diffusion junction, the base material — usually an n-type germanium or silicon — is heated to a high temperature in an enclosed vessel together with the p-type impurity. At about 1100° to 1300° C, the impurity forms sufficient vapor to begin to diffuse slowly into the surface of the base material, forming a p-n junction on each side. The n-type base material is slowly converted into p-type material as the diffusion process continues. Time and temperature control permits the formation of large flat junctions with precisely governed thicknesses. Such large junctions are necessary in power transistors where the heat generated by the losses within the transistor must be dissipated quickly. In most modern manufacturing designs, the transistor body is brought into intimate contact with a relatively large mass of copper called a *heat sink*. A heat sink conducts heat away from the transistor rapidly, keeping its temperature within tolerable limits.

ATOMS AND SEMICONDUCTORS

Current Flow Through a P-N Junction

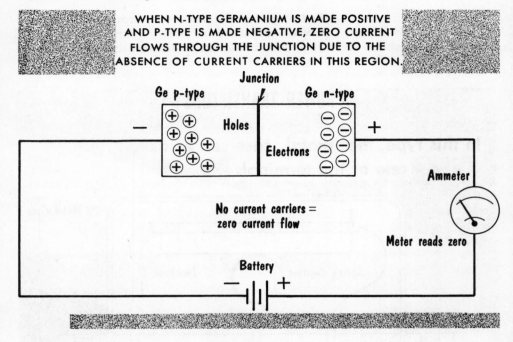

WHEN N-TYPE GERMANIUM IS MADE POSITIVE AND P-TYPE IS MADE NEGATIVE, ZERO CURRENT FLOWS THROUGH THE JUNCTION DUE TO THE ABSENCE OF CURRENT CARRIERS IN THIS REGION.

To have a current through any conductor, we must first apply a voltage. Suppose we apply a difference of potential across the ends of two pieces of transistor germanium, one p-type and the other n-type with a good, formed junction between them. Assume, to start, that we connect the negative terminal of the battery to the p-type section, and the positive terminal to the n-type section.

For this connection, we should find little or no current flowing through the bar. If we remember that the n-type germanium has free electrons (donor impurity) to act as current carriers and that the p-type has holes (acceptor) to carry current, we can see at once why the germanium offers a very high resistance to the flow of current.

The voltage source establishes an electric field within the semiconductor. This field has a direction such as to drive the electrons close to the positive terminal connection and the holes over to the place where the negative terminal of the battery is connected. This leaves no current carriers in or near the junction. Without carriers, an electric current simply cannot flow since the very word *"current"* signifies a movement of charged particles. Thus, with the n-type material connected to the positive end of the potential source and the p-type connected to the negative side, the semiconductor becomes a nonconductor.

Current Flow Through a P-N Junction (contd.)

Now let us reverse the battery connection to the n-type and p-type germanium sections without making any other changes in the circuit. We find at once that the current flowing may be very great. To protect the delicate germanium wafer and our meter, we insert a resistor in series with the other components to hold the current down to a reasonable value. What change has occurred that has resulted in converting the germanium into a conductor rather than an insulator, as it appeared to be previously?

With the reversal of polarity, the electrons and holes are driven in opposite directions; this time *right to the junction*. Under the impelling force of the electric field, they cross the junction and mingle with each other. But as an electron meets a hole, *both* charges disappear by cancellation, leaving the p-type material deficient in hole content and the n-type germanium section deficient in electron content. Each has lost one of its current carriers due to the cancellation. With the battery connected as it is, electrons can easily enter the n-type germanium to supply what it has lost

WHEN THE BATTERY IS REVERSED, BOTH ELECTRONS AND HOLES CROSS THE JUNCTION, JOIN WITH EACH OTHER, AND CANCEL. THIS LEAVES THE WAY OPEN FOR ELECTRONS TO ENTER THE N SIDE FROM THE BATTERY AND ELECTRONS TO LEAVE THE P SIDE TO THE BATTERY, THUS SETTING UP AN ELECTRON CURRENT IN THE EXTERNAL CIRCUIT.

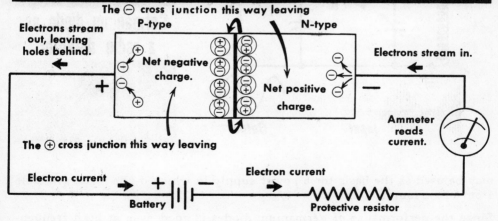

by cancellation. In like manner, the p-type material, which has a net negative charge due to the loss of positive carriers, gives up an electron to the positive end of the battery, thus forming a new hole to replace the one it lost across the junction.

Junction Rectification

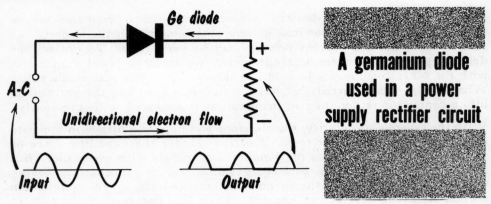

A germanium diode used in a power supply rectifier circuit

The resemblance of this behavior to that of a vacuum tube rectifier is unmistakable and at once suggests that a p-n junction device could serve as a good rectifier. All we have to do is to replace the battery with an a-c generator, retaining the protective resistor to serve as an output load resistor as in any vacuum tube circuit. Thus, a germanium junction rectifier

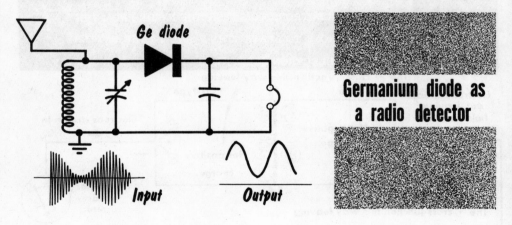

Germanium diode as a radio detector

may be used as the basis for a power supply in which d-c is made available from an a-c source.

Since the performance of germanium diodes is good even at high frequencies, they are very often used as detectors in radio circuits. In either of these applications, they offer the following advantages: (1) No warm up time required; (2) No filament power necessary; (3) Occupy very little space compared to a vacuum tube rectifier and (4), Produce little heat, simplifying the ventilation problems common to vacuum tube equipment.

Germanium Rectifier Curve

We have spoken of the absence of current when the n-type germanium is connected to the positive battery terminal and the p-type to the negative terminal. If we inserted a very sensitive microammeter in the circuit, we should find a very tiny reverse current flowing.

We account for this current by noting that in n-type germanium, where the important current carriers are electrons, there must also be a few isolated

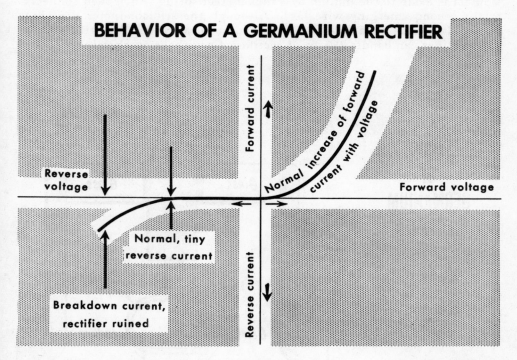

BEHAVIOR OF A GERMANIUM RECTIFIER

Forward current

Normal increase of forward current with voltage

Reverse voltage

Forward voltage

Normal, tiny reverse current

Reverse current

Breakdown current, rectifier ruined

and scattered holes. Similarly, in p-type material there must be some stray electrons. It has become conventional to call the important carriers in either material the *"majority"* carriers and the stray, scattered charges of opposite sign the *"minority"* carriers. Thus, in n-type material the *majority* carriers are the *electrons* and the *minority* carriers are the *holes*. *Electrons* are the *minority* carriers in p-type semiconductors, with *holes* acting as the majority carriers. Thus, when the voltage is applied in the resistance direction, the few microamperes that flow may be attributed to the minority carriers in both sections. As this reverse voltage is increased, the minority carrier current increases slightly until a breakdown point is reached. When this happens, the current increases suddenly, indicating a destruction of the crystalline structure in the junction and usually, the end of the usefulness of that particular germanium rectifier.

Germanium and Silicon Rectifiers

The actual structures of germanium and silicon rectifiers are interesting to examine. They are almost identical in outward appearance but differ considerably in internal structure. In the former, a tiny germanium pellet is soldered to the base metal from which the connecting lead is brought out. This metal is generally brass or a similarly suitable alloy. The indium bead used to form the junction is immediately above the germanium pellet. Contact is made to the indium by a nickel-steel spring which then connects to the second emerging wire lead. Although the semiconductor does not exceed 1/16 inch on a side while the indium bead is smaller still, such a tiny rectifier can handle 250 ma of rectified d-c without overheating.

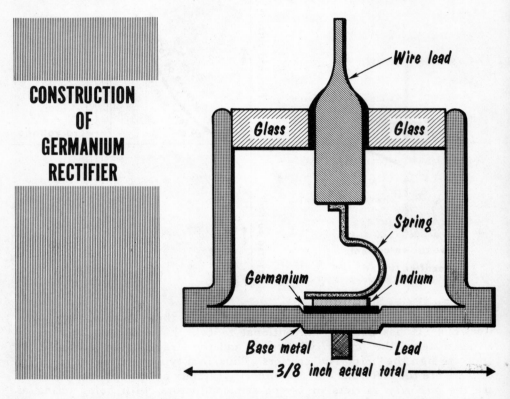

CONSTRUCTION
OF
GERMANIUM
RECTIFIER

Wire lead

Glass   Glass

Spring

Germanium   Indium

Base metal   Lead

3/8 inch actual total

When ambient temperatures rise above 85°C., the silicon rectifier is preferred. In this unit, a small piece of pure silicon — again about 1/16 inch on each side — is secured to the base where contact is made with it. A short length of aluminum wire is alloyed with the silicon to form a p-n junction. This rectifier can handle up to 750 ma at room temperature without difficulty and will operate satisfactorily at ambient temperatures much higher than the germanium type.

Important Semiconductor Rectifier Ratings

RATINGS OF TWO SEMICONDUCTOR RECTIFIERS	LOW CURRENT GERMANIUM TYPE 1N91	HIGH CURRENT SILICON TYPE 4JA60
Peak inverse voltage	100	300
D-C output current (The useful output current as delivered to a load operated by the power supply)	150 ma	100 amps (if temperature is held below 100°C)
D-C surge current (Usually limited to a single cycle of the a-c input)	25 amps	900 amps
Full-load voltage drop (The voltage lost to the load due to the internal drop in the rectifier. Note how small this is)	0.5	1.0 (approx. at 100 amperes)
Leakage or reverse current at 50 volts reverse bias	125 μa	50.0 ma
Operating frequency for approx. 70% rectifier efficiency	50 kc	1 kc
Storage temperature (maximum)	85° C	200° C

When we discussed the germanium rectifier curve, we indicated that there was a certain reverse voltage beyond which the rectifier would break down and that exceeding this voltage would result in the destruction of the rectifier unit. This voltage level is called the *maximum inverse peak voltage rating* of the rectifier. It corresponds quite closely to the rating of the same name given to vacuum tube rectifiers.

Other important ratings essential for selecting a particular rectifier for a given job are listed in the chart. To indicate the wide range of operating voltages, currents, and temperatures, we have given the ratings for a low-current germanium type as previously described and a high-current silicon rectifier.

Simple Power Supply for Transistorized Devices

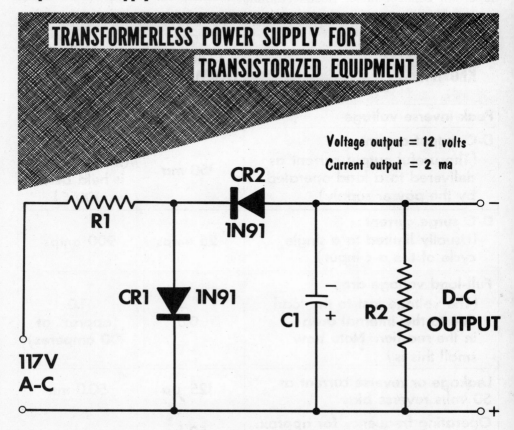

TRANSFORMERLESS POWER SUPPLY FOR
TRANSISTORIZED EQUIPMENT

Voltage output = 12 volts
Current output = 2 ma

Many transistorized devices, such as preamplifiers for high fidelity systems, sensitive relays, and r-f amplifiers, can be equipped with a very simple power supply that will provide 12 volts at 1 or 2 ma. Such a power supply can literally be built into a match box since it requires no transformer or other large parts.

To keep the size down, a tiny germanium rectifier must be used. But as the chart in the previous section shows, the peak inverse voltage rating of a representative rectifier, such as the 1N91, is only 100 volts. This is far too low to permit using the 1N91 as a simple half-wave rectifier on a 117-volt a-c line. How can we protect the rectifier against reverse voltage breakdown and still avoid the use of a transformer? The solution is shown in the diagram. A second 1N91 is connected in series with a large resistor, R1, and is shunted across the a-c line. The best way to explain the operation of the power supply is to break it down into several parts and examine the voltages and currents in each part for both halves of the a-c cycle.

How the Power Supply Works

We begin by assuming the input voltage polarity to the top line as plus, and to the bottom line as minus. For this polarity, CR2 does not conduct (except for a negligible reverse current) while CR1 is fully conducting. Since the forward resistance of CR1 is negligibly low, virtually the entire line voltage of 117 volts appears across R1, while the drop across CR1 is

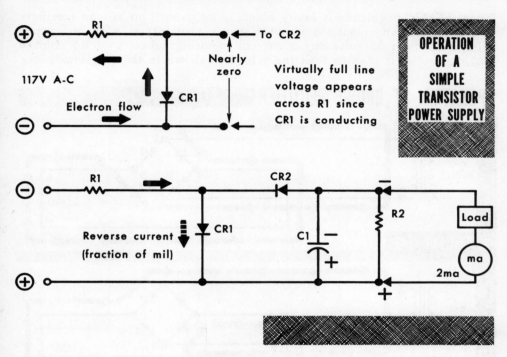

close to zero. Thus, for this half-cycle there is a very small voltage across CR1 and a small one across CR2 due to the charge on the capacitor C1. This charge is obtained on the alternate a-c cycle and is close to 12 volts. No danger of exceeding the maximum peak inverse voltage here.

CR1 becomes nonconducting except for a reverse current of a fraction of a milliampere when the a-c polarity reverses. Under this condition the inverse voltage of CR1 is of importance. At the same time CR2 begins to conduct and to supply current to the load, resulting in a circuit which consists of the parallel network R2, reactance of C1, and reverse-resistance of CR1, in series with R1. C1 charges to supply the current for the load on the other alternate cycle. Because of the large value of C1, its reactance is much smaller than R2 and reverse-resistance of CR1. Thus the voltage across C1 (and hence CR1) is kept small in relation to the voltage across R1. This is more than adequate to insure that the peak inverse voltage that appears across CR1 is far below the maximum rating.

High-Current Bridge Rectifier Systems

The power supply just described is more than adequate for low-current applications since the approximate ripple voltage it produces is only 0.1% of the d-c voltage supplied to the load. When a power supply for power transistors is required, as we shall see later, we shall want voltages in the order of 30 volts and currents as high as 300 or 400 ma.

Power of this magnitude is easily obtained by using four silicon rectifiers in a bridge arrangement. A suitable power transformer with an output voltage of about 25 volts and a current handling capacity of 0.5 ampere (500 ma) or so, is used to feed the bridge as shown in the diagram. Since

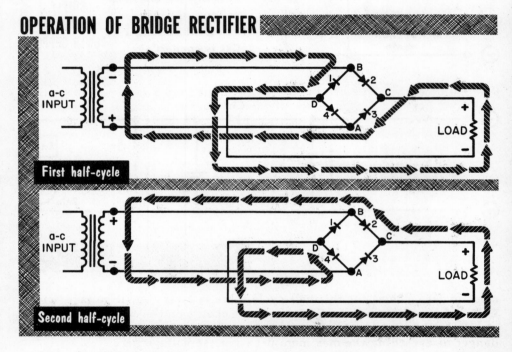

OPERATION OF BRIDGE RECTIFIER

a bridge provides full-wave rectification, a single filter capacitor is all that is needed to smooth the output sufficiently for power amplification. (Power amplifiers have little *voltage* amplification and therefore are insensitive to ripple voltages in their power sources that might produce hum in a voltage amplifier.)

The operation of the power supply is identical to that of a vacuum tube bridge. When point *A* goes positive with respect to *B*, conduction occurs through rectifiers 1 and 3 with the electron current going through the load in the direction shown. When *B* is positive with respect to *A*, the sequence of conduction is: through 4, through load, through 2, and back to *B*.

The Point-Contact Diode

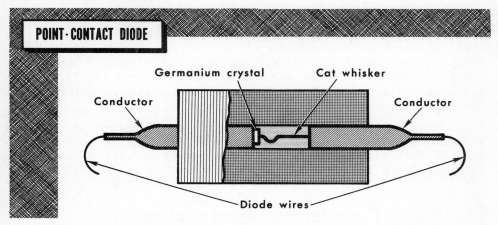

POINT-CONTACT DIODE

Germanium crystal

Cat whisker

Conductor

Conductor

Diode wires

The point-contact type of semiconductor diode varies from the previously described junction diodes in the method of construction and other characteristics. In the construction of a point-contact diode, a piece of n-type germanium is metal coated on one side and forms one of the diode terminals. A fine phosphor-bronze or beryllium-copper wire, often called a cat whisker, is pressed against the center of this n-type germanium slab. The other end of this cat whisker forms the other terminal of the diode.

To create the p-type material, a rather large current is passed through the cat whisker to the germanium slab. During the short instant that the current flows, a small "island" or dot of p-type germanium is formed directly under the point where the wire touches the crystal slab. Thus the point-contact diode is actually a junction diode where one of the semiconductor materials is of extremely small area. The explanation of the

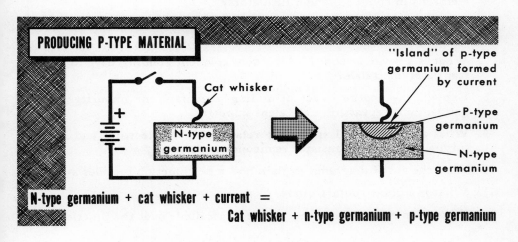

PRODUCING P-TYPE MATERIAL

Cat whisker

N-type germanium

"Island" of p-type germanium formed by current

P-type germanium

N-type germanium

N-type germanium + cat whisker + current =

Cat whisker + n-type germanium + p-type germanium

operation of the point-contact diode proceeds along lines similar to the junction diodes.

Point-contact diodes are low-power devices. Their chief advantage over junction diodes is the small area of the junction. Because of this small area, the capacitance across the junction is much smaller than for the junction diode. Hence, the point-contact diode is used mostly in video second detectors and related applications where advantage is taken of its low junction capacitance.

There is not much difference between germanium and silicon point-contact diodes. The latter is often used as a microwave mixer. In general, point-contact diodes are more fragile and will withstand less reverse voltage than junction diodes.

QUESTIONS AND PROBLEMS

1. State the advantages of transistors as compared with vacuum tubes in the processes of amplification and oscillation.

2. What is meant by a diatomic molecule? Give an example of a diatomic molecule.

3. Name the three fundamental atomic particles, stating the polarity of their electric charges.

4. Although the helium atom contains only two protons, its mass is approximately four times the mass of one proton. Explain how this is possible.

5. What is a valence electron? What is the importance of the valence electrons in conductors and insulators?

6. Silicon and germanium are normally nonconductors. How are they made into semiconductors in transistors?

7. What is a covalent bond? What does covalent bonding have to do with transistor materials?

8. Explain the charge effect of adding a pentavalent impurity to pure germanium; repeat for a trivalent impurity.

9. What are "holes"? Explain the relationship of electrons and holes to p-type and n-type transistor semiconductors.

10. With the aid of diagrams, explain how a p-n junction rectifies a-c.

11. What is a point-contact diode?

12. What is an advantage of the point contact diode over the junction type?

From Germanium Diodes to Transistors

The transistor is an outgrowth of experimental work on germanium rectifiers, although nine years elapsed between the publication of details con-

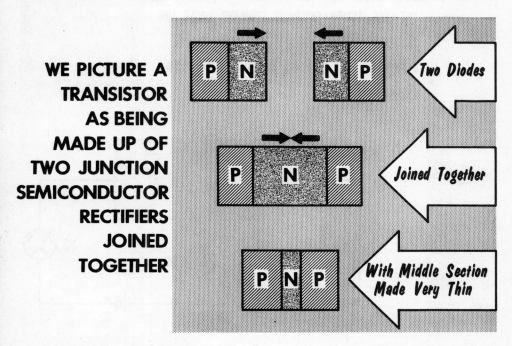

WE PICTURE A TRANSISTOR AS BEING MADE UP OF TWO JUNCTION SEMICONDUCTOR RECTIFIERS JOINED TOGETHER

cerning the junction semiconductor rectifier and the announcement of the first practical transistor.

It will be easier to transfer your understanding of germanium rectifier action to transistors, by imagining that a transistor is made by bringing two diodes together. Holding the two diodes with like transistor materials facing each other, we bring them together so that the like materials fuse with no junction between them. We then squeeze them together until the fused transistor semiconductor becomes a very thin slice sandwiched between two much larger pieces of the opposite kind of material. In the illustration, we show the fused material as n-type with the "bread" of the sandwich of p-type material. This arrangement is called a p-n-p transistor. By starting with fused p-material and n-type germanium as the outside sections, we can thus fabricate an n-p-n type of transistor.

Of course, transistors are not made this way. We have already studied the fused-alloy and diffused-junction methods of manufacture. It is valuable at this time, however, to visualize a transistor as being composed of a pair of fused rectifiers.

Applying Bias to a Transistor

BIASING A TRANSISTOR

THE LEFT SECTION (N1-P) IS FORWARD BIASED AND THE RIGHT SECTION (N2-P) IS REVERSE BIASED. R PROTECTS THE FORWARD BIASED SECTION FROM EXCESSIVE CURRENT

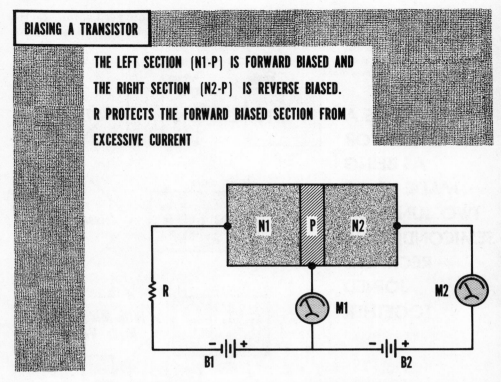

To help us in the fundamental understanding involved in transistor action, let us see what happens when we apply some voltages to the n-p-n structure. Considering the left-hand diode section made up of n1 and p first, we have connected a battery in such a way that conduction can occur through the diode. To refresh your memory, you can refer back to the discussion of the current flow through p-n junction. Resistor R is placed in the circuit to prevent excessive current flow. The electric field has the right polarity to make electrons and holes recombine at the junction between n1 and p. Thus, electrons will flow from the negative end of B1, through n1 to p, out of p, through meter M1, and back to the + end of B1. This condition is known as *forward bias* in a transistor because the battery connection is such as to produce a low resistance at the n1-to-p junction.

Analyzing the right side of the transistor, we see that the battery is connected so that the p-to-n2 junction has a very high resistance as in the case of a diode rectifier connected for nonconduction. In this case, battery B2 supplies very little current to the p-n2 section and meter M2 reads only a few microamperes without any sort of protective resistor in the circuit. This connection is referred to as *reverse bias*.

Effect of a Thin Central Section

The preceding analysis of forward and reverse bias is based upon the assumption that all of the transistor sections are relatively thick. If a transistor were made this way, it simply would not work very well. It would merely be a pair of diodes — certainly not a device that can amplify, oscillate, perform switching functions and a host of other tasks.

Let us now thin down the central section to less than 1/1000 inch. In practical transistors, the central p-slice is made of high-purity germanium so that it is only slightly p in nature. Thus the n1 section has lots of free electrons, but the thin central p-section has very few holes. Consequently, the major carriers of electricity are electrons.

With the application of battery voltage from B1, electrons begin to diffuse across the n1-p junction in large numbers. They find very few holes with which they can recombine. The p-section cannot become very negative due to loss of holes, since it does not have many holes to start. Consequently, there are so few electrons freed from the p-material that only a very small number return via M1 to the postive end of B1. Even when R is removed, we find this current to be quite small, making it possible to omit the protective resistor altogether. We see that even with forward bias, very few of the electrons that enter the p-section from n1 can find their way back to the positive terminal of B1. In essence, the base region (due to its low doping) is a thin high resistance material. What happens to all the electrons that diffused across the n1-p junction?

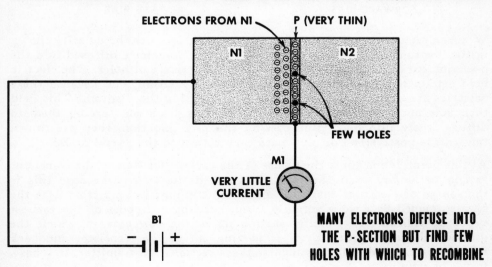

EVEN WITH A FORWARD BIAS, THE CURRENT FROM B1 IS VERY SMALL DUE TO THE SCARCITY OF HOLES IN THE VERY THIN P-SECTION

ELECTRONS FROM N1   P (VERY THIN)

N1   N2

FEW HOLES

M1

VERY LITTLE CURRENT

B1

MANY ELECTRONS DIFFUSE INTO THE P-SECTION BUT FIND FEW HOLES WITH WHICH TO RECOMBINE

Effect of the Second Battery

THE PARTS OF A TRANSISTOR AND ITS CIRCUIT
(n-p-n)

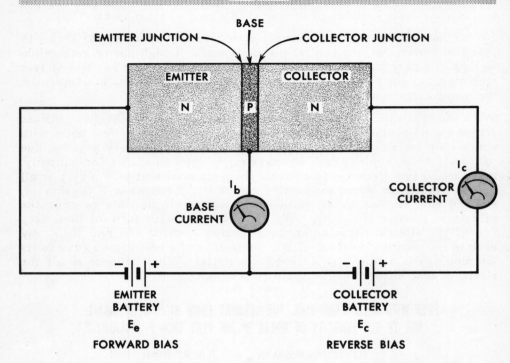

When another battery is connected to the transistor with the polarity shown in the diagram, a new effect occurs. Many electrons have diffused into the p-section but cannot be neutralized due to the scarcity of holes. The electric field set up by both B1 and B2 is a relatively strong one because these batteries are in series as may be seen by inspecting the diagram. This field then acts upon the free electrons in the thin p-section, forcing them to diffuse freely into n2. After passing the p-n2 junction, they are drawn toward the postive end of n2, where they return to the + end of B2.

At this point, let us apply the names of the various sections of the transistor as conventionally used. Section n1 is called the *emitter* because this is the source of electrons that diffuses across the junctions; section p is the *base* so named because it actually formed the base support of the earliest types of transistors. Finally, section n2 is the *collector* in which the diffused electrons gather prior to moving out of the transistor and back into the battery. We shall symbolize these sections: E = emitter, B = base, C = collector.

Base Current and Collector Current

If the emitter battery is disconnected from the emitter with no other changes made in the arrangement just described, a *very tiny* current flows in the collector circuit as indicated by the I_c meter. This current is due to the presence of a few minority carriers in the base and collector. A large current cannot flow because the collector is reverse biased. The tiny leakage current is generally symbolized by I_{CBO}, meaning *collector current with emitter circuit open*.

Now let us connect the batteries. Let us further assume that the voltage and resistance conditions are such that 10 ma flow out of the negative side of the emitter battery into the emitter. As these electrons are impelled

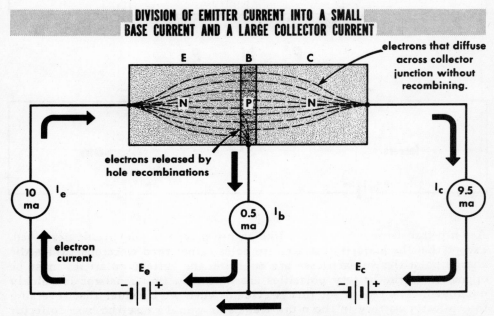

DIVISION OF EMITTER CURRENT INTO A SMALL BASE CURRENT AND A LARGE COLLECTOR CURRENT

The greater the difference of potential between E and B, the greater becomes I_c. Thus, the base controls the flow of collector current, without taking much of the initial current itself.

across the emitter-base junction, about 5% of the 10 ma will be represented by electrons that recombine with holes in the base, producing a base current (I_b) of 0.5 ma. The remainder of the current — 9.5 ma in this case — will be carried by electrons that diffuse across the collector junction and end up as collector current, I_c. From these two situations and others that follow, we *conclude* that the emitter battery controls the potential between emitter and base, that this potential controls the current that flows from emitter to collector, but *that the base itself takes very little of this current.*

Rules for Biasing N-P-N and P-N-P Transistors

ILLUSTRATING THE RULES FOR TRANSISTOR BIASING

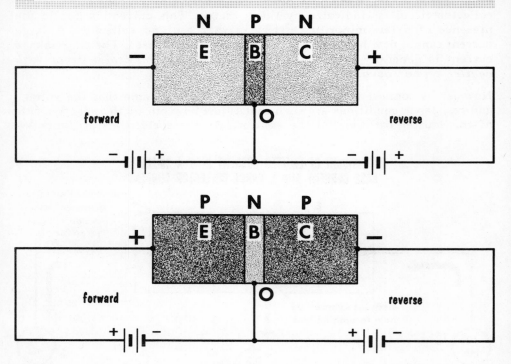

A p-n-p transistor behaves just like an n-p-n type in the circuit described, except that the majority carriers are holes rather than electrons. Since the charge polarities in each case are opposite, the battery polarities must be opposite. If the battery polarities are incorrect, the transistor is quickly destroyed. The reason for this is evident when we consider that reversing the collector battery in the n-p-n transistor would cause the base-collector circuit to be *forward biased*. The heavy current thus caused through the base-collector junction would overheat and destroy the transistor in a matter of seconds. This gives us one general rule and two specific rules to remember about biasing transistors.

General Rule: The emitter-to-base circuit of any transistor is always forward biased; the base-to-collector circuit is always reverse biased.

Specific Rules: For n-p-n transistors, the collector must be positive with respect to the base, while the emitter must be negative with respect to the base. For p-n-p transistors, the collector must be negative with respect to the base, while the emitter must be positive with respect to the base.

Comparing a Transistor to a Triode

Although the operation of a transistor is very different in theory from that of a vacuum tube, certain similarities exist. Very often such comparisons are valuable because they permit one to visualize the action of an unfamiliar device in terms of a more familiar one.

TRANSISTORS AND TRIODES COMPARED

Similarities		Why They Are Similar
VACUUM TUBE	TRANSISTOR	
Cathode	Emitter	Both electrodes act as *sources* of current carriers.
Grid	Base	Both electrodes *control* larger amounts of power than they dissipate in themselves.
Plate	Collector	Both electrodes *capture* the charge carriers that ultimately supply the output power.

We must not ignore the significant differences between vacuum tubes and transistors, however. For instance, there is only *one* kind of triode, but there are *two* kinds of transistors — n-p-n and p-n-p. In triodes, the only kind of charge carrier is the electron. In transistors, both electrons and holes are utilized as charge carriers. This gives us one specific rule for biasing triodes: the grid is more negative than the cathode while the plate is more positive than the cathode.

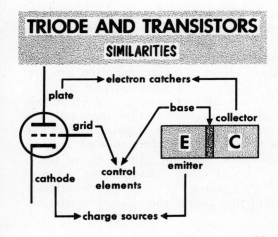

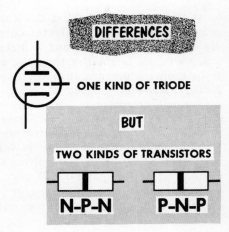

The Meaning of Alpha

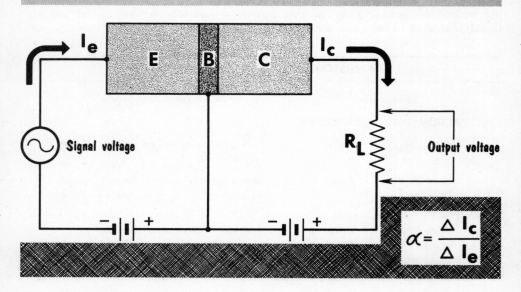

THE MEANING OF ALPHA

$$\alpha = \frac{\Delta I_c}{\Delta I_e}$$

We have seen that all of the electrons that enter the emitter from the battery do not find their way into the collector circuit. Some of the diffusing electrons in the base region recombine with the few holes that are there, causing a small base current to flow. Since we are interested in the transistor as an *amplifier*, we shall first investigate what happens when a source of signal voltage (a-c) is introduced into the input circuit of the transistor. In addition, let us connect a load resistor in series with the collector battery to act as an output load across which the output voltage can develop.

As the signal source varies the emitter bias voltage — and, of course, the emitter current — the collector current varies at the same frequency. When the extent of the variations is measured, it is found that the collector current variation is *smaller* than the emitter current variation due to the loss of incoming electrons into the base circuit by the recombination process. Thus, instead of getting a current gain, we get a small *current loss*. In an average commercial transistor, the current loss is about 5% since this is the percentage of recombination to be expected. Rather than call it a current loss of 5%, we prefer to say that we have a *current gain* of 0.95. The fraction shows that the gain is less than 1. The performance of a given transistor in terms of its emitter-to-collector current gain is called the α (alpha) of the transistor. The definition of alpha follows from the preceding discussion and may be given as: the ratio of a change (Δ) in collector current to the change in emitter current that caused the collector variation.

Transistor Resistances

The general rule previously given for biasing transistors indicates that the emitter-to-base circuit of a transistor is always forward biased, while the base-to-collector circuit is always reverse biased. Another way to state this is the emitter-to-base circuit is biased in the *low resistance* direction, and the base-to-collector circuit is biased in the *high resistance* direction. What can this tell us about the battery voltages that can be used in each of these circuits?

The battery in the emitter-to-base circuit must be a very low voltage type. As we shall see later, only 0.1 to 0.5 volt are necessary to produce a sizeable emitter current through this low resistance. In the base-to-collector circuit, however, we can use a much higher battery potential because this circuit has a *high* resistance. The actual voltage used for E_c depends upon the design of the individual transistor. For many modern types, this voltage may be as high as 45 volts.

Although the actual forward resistance of the emitter circuit and the reverse resistance of the collector circuit naturally varies from one transistor to another, we will select representative figures around which we can weave our concepts. For the moment, we shall take the typical emitter-to-base resistance as about 250 ohms, and the typical base-to-collector resistance as 300,000 ohms.

BATTERY ARRANGEMENT FOR TYPICAL TRANSISTOR

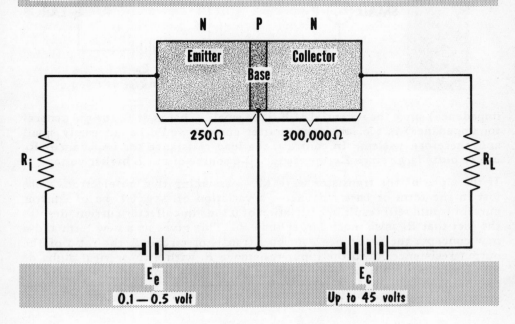

Resistance Gain

Imagine a signal generator connected in the emitter circuit in place of R_i. Such a generator would have to have a low internal resistance (or

> Assuming an alpha of 1.0, the same variations in collector current occur as there are variations in emitter current, despite the fact that R_i is small and R_L is large. The ratio of R_L to R_i is called the resistance gain of the circuit.

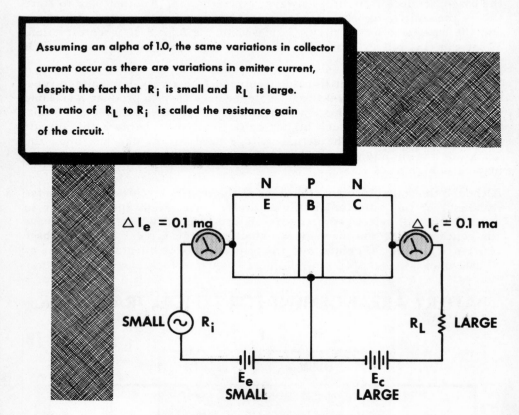

impedance) since the potential of E_e is so small. That is, if R_i (or the generator impedance) is too large, the emitter current would be extremely small and, therefore, useless. In contrast, the load resistance (or impedance) R_L *can be quite large* since E_c represents a d-c source of much higher voltage.

If the alpha of the transistor were 1.0 — assuming that no electrons were lost in the form of base current — a variation of, say, 0.1 ma of emitter current would still result in a variation of 0.1 ma in collector current, despite the fact that R_L is so much larger than R_i. This gives us a new term and a new concept: the *resistance gain* of a transistor circuit is the ratio of the output resistance R_L to the input resistance R_i with a theoretical alpha of 1.0. Resistance gain merely expresses how much more output resistance can be used as compared with input resistance for essentially the same emitter and collector currents.

Voltage Gain of a Transistor

Now we come to considerations that show us how and why a transistor can act as an amplifier. As in the case of vacuum tubes, amplification involves *voltage gain* which is defined as follows:

$$V\ G = \frac{\Delta\ E_L}{\Delta\ E_i}$$

E_L is voltage *output* while E_i is voltage *input*. Voltage gain is the ratio of a change of output voltage to the change in voltage input. From Ohm's Law, we know that $E = IR$. Thus, replacing the E's above with the corresponding IR's, we can write:

$$V\ G = \frac{\Delta\ I_c R_L}{\Delta\ I_e R_i}$$

As we have seen previously, however, the ratio of $\Delta I_c/\Delta I_e$ is the current gain of the transistor. Therefore, we can substitute alpha for this ratio in the equation above:

$$V\ G = \alpha\ \frac{R_L}{R_i}\ .$$

From this it is evident that if R_L is large compared to R_i, the ratio of the two resistances will give us a large number by which to multiply alpha. For instance, a common generator resistance is about 100 ohms while the load resistance might be 100,000 ohms. If alpha is 0.95 or so, then the voltage gain is $0.95\ \times\ \dfrac{100,000}{100}$ or:

$$V\ G = 0.95\ \times\ 1000 = 950.$$

VOLTAGE GAIN IS THE PRODUCT OF CURRENT GAIN AND RESISTANCE GAIN

Power Gain in Transistors

A measure often used to evaluate the performance of an amplifier is that of *power gain*. This factor can be described as the ability of an amplifier to *control* large amounts of output power with small amounts of input

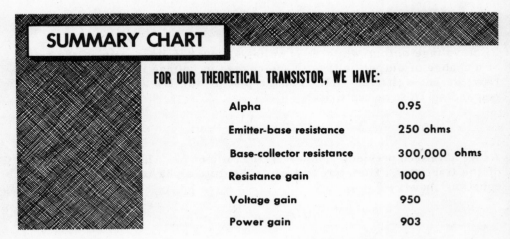

SUMMARY CHART

FOR OUR THEORETICAL TRANSISTOR, WE HAVE:

Alpha	0.95
Emitter-base resistance	250 ohms
Base-collector resistance	300,000 ohms
Resistance gain	1000
Voltage gain	950
Power gain	903

power. On this basis, power gain may be defined as the ratio of a change of output power to a change of input power that produced it: $P\ G = \dfrac{\Delta P_o}{\Delta P_i}$ From Ohm's Law, power may be found by multiplying the square of the current by the resistance through which the current flows, or $P = I^2R$. And since power output is *collector load* power while input power is the power supplied by the generator to the emitter circuit, we may rewrite the first equation this way:

$$PG = \frac{\Delta I_c^2 \times R_L}{\Delta I_e^2 \times R_i}.$$

The first term in this fraction is α^2, so that the final equation becomes

$$PG = \alpha^2 \times \frac{R_L}{R_i}$$

Again assuming alpha to be about 0.95, and using the same input and output resistances as before, we can determine the power gain of our theoretical case by substituting these values in the final equation as follows:

$$PG = (0.95)^2 \times \frac{100,000}{100} = 903.$$

This is a large power gain. It shows that transistors are capable of providing power gains comparable to that of tubes. Furthermore, in practice it is possible to realize power gains up to 40,000 in some special applications.

Symbolization

In the preceding pages, we have drawn transistors in *pictorial* form because the movement of the charge carriers can be seen more easily when the emitter, base, and collector are laid out in a straight line. As our circuits develop, however, this kind of drawing becomes unwieldy and difficult to follow. In the few years transistors have been in existence, their symbols have become surprisingly standardized. Use of these symbols not only permits you to understand drawings from any source, but it also simplifies the process of circuit pictorialization.

The fundamental n-p-n and p-n-p amplifier circuits are shown in the accompanying drawings. Note the difference between the symbols for each type of transistor. In the n-p-n transistor, the emitter is shown as an arrow directed *away* from the base; in the p-n-p variety, the emitter arrow is directed *toward* the base. Otherwise, the symbols are identical.

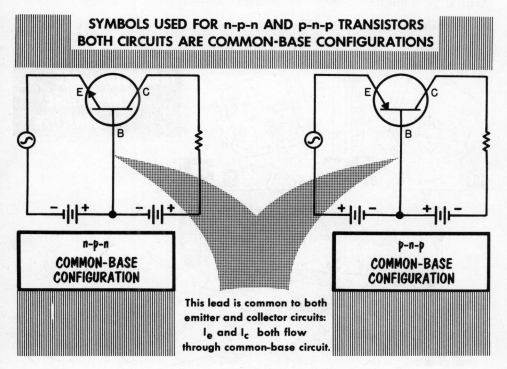

SYMBOLS USED FOR n-p-n AND p-n-p TRANSISTORS
BOTH CIRCUITS ARE COMMON-BASE CONFIGURATIONS

n-p-n
COMMON-BASE CONFIGURATION

p-n-p
COMMON-BASE CONFIGURATION

This lead is common to both emitter and collector circuits: I_e and I_c both flow through common-base circuit.

These amplifier circuits are known as *common-base configurations*. As you study the circuits, you will observe that the base electrode is common to both the emitter and collector circuits. This is one of three possible ways of connecting a transistor in an amplifier circuit. It is sometimes referred to as a *grounded base* connection. We will see the reason for this later on.

BASIC TRANSISTOR ELECTRONICS

Physical Characteristics of Transistors

Aside from their small size, there are certain physical characteristics that transistors possess that should be mentioned before we discuss their electrical characteristics.

The power handling ability of any amplifier depends upon how hot it can become without destroying itself.

A typical germanium transistor such as the 2N189 audio amplifier is rated thus:

Temperature —	Operating range	−55 to +65°C
	Storage range	−55 to +85°C

When we compare these ratings with those of the newer silicon transistors (e.g., the type 2N332) we find that these ranges have been extended substantially. Thus:

Temperature —	Operating range	−55 to 175°C
	Storage range	−65 to 200°C

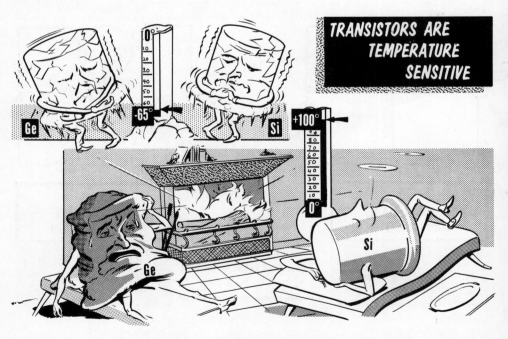

Silicon transistors, therefore, can handle more power than germanium transistors, size-for-size. In either case, exceeding the maximum ratings will almost inevitably result in the destruction of the transistor, so that much care and caution must be exercised in designing and setting up transistor circuits.

Physical Characteristics of Transistors (contd.)

A transistor can handle more power if it can transfer the heat it generates within its own body to its surroundings. Its power rating, therefore, will depend to a large degree upon *how fast* heat conduction, convection, and radiation occur.

Air is a very poor conductor of heat. Yet conduction is a very important heat transfer method. To get around this problem, many transistors are

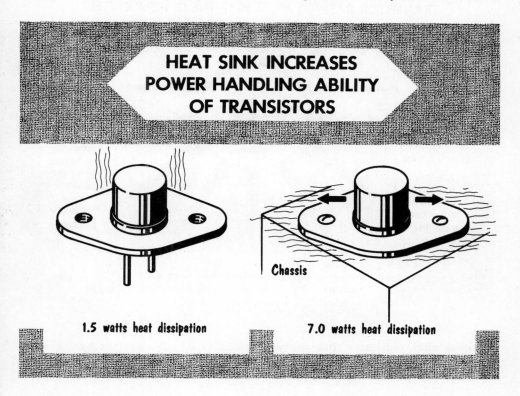

HEAT SINK INCREASES POWER HANDLING ABILITY OF TRANSISTORS

Chassis

1.5 watts heat dissipation

7.0 watts heat dissipation

constructed in such a way as to allow them to come into intimate contact with a good conductor of heat, such as a metal chassis or supporting bracket. As heat is generated in the transistor, the metal mass carries a definite percentage of it away, permitting more power to appear in the transistor without damaging it. Such a heat conductor is known as a *heat sink*. For example, a type 2N307 transistor is made so that its collector (where most of the heat is generated) is in close contact with the flat bottom of the transistor case. In free air, the 2N307 can dissipate about 1.5 watts safely but when provided with a good heat sink, it does not even begin to overheat with a dissipation of 7 watts. A larger transistor case also helps to dissipate heat by radiation into the surrounding air.

Electrical Characteristics (General Considerations)

Although there has been little standardization among various manufacturers of transistors insofar as rating charts are concerned, there are a number of individual rating factors that are common to all of them. We will discuss these in the pages that follow, selecting a modern, popular transistor type as an example, the 2N78. For the present, we shall confine ourselves

TRANSISTOR RATINGS MUST HELP THE DESIGNER SELECT THE RIGHT UNIT FOR HIS PARTICULAR PURPOSE

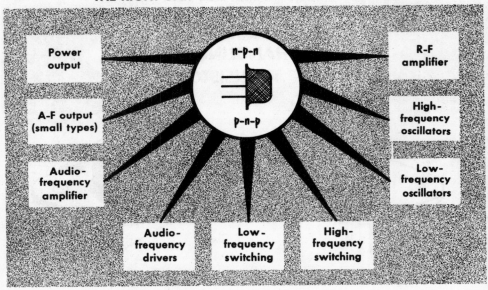

to those characteristics that apply specifically to the common base circuit. Later, we shall probe more deeply into the variations that these ratings undergo as the circuit is changed.

In general, we can divide the characteristics into three categories:

1. *Absolute maximum ratings*. These characteristics inform the user as to maximum currents, voltage, power, temperature, etc.

2. *Signal characteristics*. These deal with transistor performance under signal input conditions and provide information relative to frequency behavior, gain, impedances (input and output), and noise.

3. *Special characteristics*. Some transistors are designed for high frequency usage as i-f or r-f amplifiers and oscillators, others for audio and switching operations. Under special characteristics we would find information that would help in designing and using such circuits.

Absolute Maximum Ratings (Voltages)

ABSOLUTE MAXIMUM VOLTAGES FOR THE 2N78 TRANSISTOR

2N78   Rate-grown n-p-n junction transistor

absolute maximum ratings (25°C)

Voltages

Collector-to-Emitter Voltage (base open)	V_{ce}	15 volts
Collector-to-Base Voltage (emitter open)	V_{cb}	15 volts

(Note: The 25°C refers to the ambient or surrounding temperature. In warmer surroundings, the maximum voltages would be somewhat smaller than those shown.)

2N78 in common-base circuit.

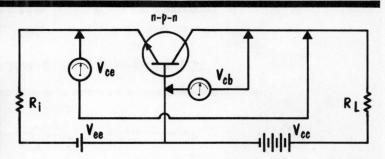

Collector-to-Base Voltage (V_{cb}) is defined as the potential applied between the collector and the base. In a d-c circuit without collector load, V_{cb} is the same as the voltage of the collector battery. From this point on, we shall adhere to the standardized symbols used in the trade. Thus, the collector battery voltage will be called V_{cc}. With a load resistance present, V_{cb} is equal to V_{cc} minus the voltage drop in the load due to collector current. For p-n-p transistors, V_{cb} is given as a negative number, indicating that the collector is minus with respect to the base of the transistor. The range of values of V_{cb} for modern transistors may be taken as approximately 6 to 80 volts.

Collector-to-Emitter Voltage (V_{cc}) rating tells us the maximum potential that can be applied between the collector and emitter without damaging the transistor. It is a less important characteristic than V_{cb}, and is not always given by the manufacturer. In general, it is not much different than V_{cb} for a given transistor.

Absolute Maximum Ratings (Maximum Currents)

Collector Current (I_c). Since all electronic components are subject to breakdown due to high temperatures generated within them by the currents they carry, the maximum current rating is an important one to observe. In general, the collector current helps us to estimate the power being dissipated in the transistor. The maximum collector current rating of a given transistor depends upon its size, its internal construction, and the manner in which heat is removed. Maximum collector currents vary widely, from a few milliamperes for small transistors to several amperes for the larger power transistor types. Collector current is given as a positive number for n-p-n transistors and as a negative number for p-n-p units.

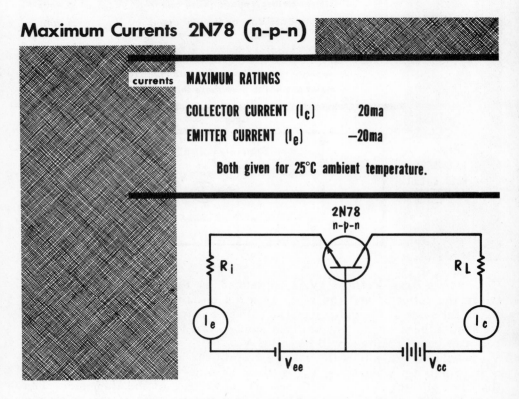

Maximum Currents 2N78 (n-p-n)

currents   **MAXIMUM RATINGS**

COLLECTOR CURRENT (I_c)   20ma

EMITTER CURRENT (I_e)   −20ma

Both given for 25°C ambient temperature.

Emitter current (I_c). The maximum emitter current of a transistor is often given along with its other maximum ratings. In most circuits, the base current is very small in comparison to the collector current. This means that the emitter and collector currents are virtually the same in value under normal operating conditions. This is also true of their maximum ratings. For n-p-n transistors, the emitter current is always shown as a negative number while for p-n-p types it is given as a positive number.

Absolute Maximum Ratings (Temperature Factors)

TEMPERATURE FACTORS 2N78 (n-p-n)

MAXIMUM TEMPERATURES AND DISSIPATION

* MAXIMUM COLLECTOR DISSIPATION (P_c)   65 mw at 25° C

MAXIMUM JUNCTION TEMPERATURE (T_j)   100°C

MAXIMUM STORAGE TEMPERATURE (T_{stg})   85°C

* Derate 1.1 mw/°C increase in ambient temperature. This means that the 2N78 may be operated with a collector power dissipation of 65 mw at an ambient temperature of 25°C, and that its maximum dissipation must be reduced (derated) 1.1 mw for each degree C above 25° C. For example, at 35°C ambient temperature, the maximum safe dissipation is 65-(1.1 x 10) = 65 - 11 = 54 mw.

Maximum collector dissipation (P_c). In the last analysis, the maximum collector dissipation rating, stated in milliwatts or watts, determines how much power input the transistor can handle. This rating is always given for some specific ambient temperature and cannot be exceeded without causing serious overheating of the collector junction. Here again, a wide span of possible ratings can occur, depending upon the design of the transistor. This maximum dissipation must also be de-rated for rises in ambient temperature as shown in the chart.

Junction Temperature (T_j). Although this maximum figure is often omitted in manufacturers' rating sheets because it is more or less constant for given semiconductor materials, it should be noted that it is about 100°C for germanium and close to 200°C for silicon.

Storage Temperature (T_{stg}). To prevent the deterioration of idle transistors, they should be stored at a temperature not exceeding the maximum rating. These range from about 85°C for germanium to 200°C for silicon.

51

Signal Characteristics

Current Gain (α). This characteristic has already been described in detail for d-c voltages and currents. When measured under signal conditions, the frequency at which the measurement is made must be given by the manufacturer. This is usually a low frequency, between 200 and 1000 cycles.

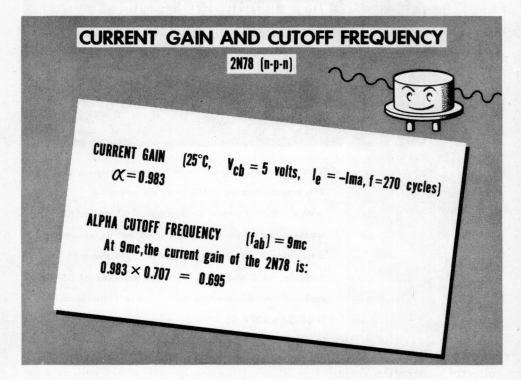

CURRENT GAIN AND CUTOFF FREQUENCY
2N78 (n-p-n)

CURRENT GAIN (25°C, V_{cb} = 5 volts, I_e = –1ma, f = 270 cycles)
α = 0.983

ALPHA CUTOFF FREQUENCY (f_{ab}) = 9mc
At 9mc, the current gain of the 2N78 is:
$0.983 \times 0.707 = 0.695$

Alpha Cutoff Frequency (f_{ab}). The current gain of a transistor falls off as the frequency increases. This drop may be attributed to many causes, chief among which are the shunting capacitance of the transistor electrodes, and the time required for electrons and holes to move through the junctions. Alpha cutoff frequency is that frequency at which the current gain as given above drops to a value that is 0.707 times as great. For instance, if the current gain of a transistor is exactly 1.0, there will be some frequency at which the gain is down 1.0×0.707. Thus, for this case, when the current gain becomes 0.707, the frequency at which this occurs is called the alpha cutoff frequency. In a practical case where alpha is equal to 0.98, the alpha cutoff frequency occurs where the current gain drops to 0.98×0.707 or 0.694. However, a transistor does not lose its usefulness at the cutoff frequency; this rating merely provides information relative to the performance to be expected from the transistor at increasing frequencies.

Signal Characteristics (contd.)

Input Impedance (Z_i). When a signal source is connected from emitter to base, a definite base current will flow. The *resistance* of this circuit is typically in the order of 250 ohms. This is not the input impedance, however, since impedance is always measured for a-c potentials or currents. Thus, we define input impedance as the load *seen* by the signal source as it works into the input circuit of the transistor. Input impedance differs not only for various transistors, but also for various circuit arrangements of the same transistor. At present, we are confining our figures to the common base arrangement. Since impedance is a frequency-sensitive characteristic, input impedance is always given with reference to a definite signal frequency.

Output impedance (Z_o). The base-collector junction of a transistor behaves like a signal generator while the transistor is in normal operation. When we speak of the output impedance of a vacuum tube amplifier, we are referring to the apparent impedance of the plate circuit *as it appears* to the

IMPEDANCES OF COMMON-BASE CONNECTION

2N78 (n-p-n)

INPUT IMPEDANCE (Z_i) ———— 55 ohms
  (270 cycle input signal)
  (output short-circuited)

OUTPUT IMPEDANCE (Z_o) ———— 5 megohms
  (270 cycle input signal)
  (input open-circuited)

Open-circuited input means open-circuited with respect to a-c. In all measurements d-c bias values are maintained at their operational value.

output load device. Similarly, the output impedance of a transistor is defined as the apparent impedance of the output circuit *as seen by the output load*. Z_o is normally higher in value than the output resistance of the same transistor.

Signal Characteristics (contd.)

Noise Figure (NF). One of the essential functions of transistors is that of amplification. Any noise generated within the transistor is amplified along with the impressed signal. This is a serious problem, particularly when

Noise Figure (NF) —— 12 db
$V_{cb} = $ 1.5 volts
$I_e = $ -0.5 ma
$f = $ 1000 cycles

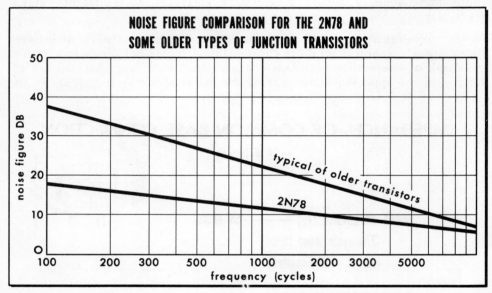

transistors are used in high-gain, high-fidelity audio amplifiers. As the noise figure is reduced, performance improves. For example, older types of transistors were rated as having noise figures in the vicinity of 60 to 80 db. (The db is a unit that may be used for measuring sound intensity. Its exact meaning need not concern us here.) Today we find transistors commercially available with noise figures under 10 db. Like other transistor characteristics noted previously, noise figure is frequency-sensitive, *increasing* in value at *lower* frequencies. Hence, when a manufacturer specifies a noise figure, he must present accompanying details such as the collector voltage used during measurement, the emitter current flowing at the time and the frequency at which the measurement was performed.

Vacuum tubes show considerable superiority with respect to noise figure. For example, a 6AG5 pentode at 1 kc has a noise figure of only 3 db, while a 6BZ7 cascode triode is rated at 2.1 db.

Signal Characteristics (contd.)

Output Capacitance (C_o). In general, the output capacitance rating of a transistor assumes increasingly greater importance as the operating frequency rises. Since C_o acts as though it were in parallel with the load, it has the effect of bypassing the higher frequencies around the load, causing high frequency losses. Thus, it is to the interest of the user that the transistor has as low an output capacitance as possible. In this respect, modern transistors compare very favorably with electron tubes.

Power Gain (PG). Power gain, like most of the other transistor characteristics, depends upon the circuit used and the frequency of operation. Since we are discussing the common base arrangement, the power gain for this configuration is given in the chart for a frequency of approximately 500 kc. This characteristic is virtually always given in decibels, since this

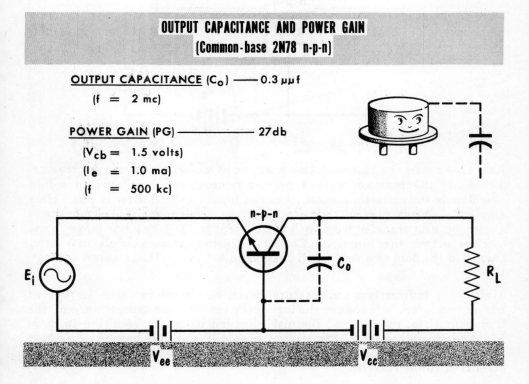

OUTPUT CAPACITANCE AND POWER GAIN
(Common-base 2N78 n-p-n)

OUTPUT CAPACITANCE (C_o) —— 0.3 μμf
 (f = 2 mc)

POWER GAIN (PG) ———————— 27 db
 (V_{cb} = 1.5 volts)
 (I_e = 1.0 ma)
 (f = 500 kc)

E_i n-p-n C_o R_L V_{ee} V_{cc}

unit has become standard for expression of power ratios. It can be converted into an actual gain ratio, however, by the following method: merely divide the number of db by 10 and find the antilog (to the base 10) of the quotient. For example, a PG of 27 is approximately equivalent to a power ratio of 520 to 1.

Collector Cutoff Current (I_{co})

COLLECTOR CUTOFF CURRENT
2N78   n-p-n

Collector Cutoff Current   (I_{co})

$V_{cb} = 15$ volts	$I_{co} =$ 5 microamperes
$V_{cb} = 5$ volts	$= 0.7$ microamperes

I_{co} is always measured with the input open circuited

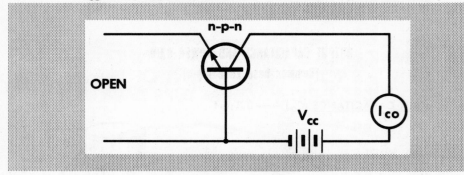

As we have seen, the base-collector junction of a transistor is always reverse-biased. If the junction were a perfect nonconductor, no current would ever flow in the collector circuit when the input to the emitter is zero. Heat energy is always present, however, and the random molecular movement caused by heat transfer from the environment to the transistor moves some carriers across the junction. Carrier diffusion, though small in extent, results in the flow of a very small current called the collector cutoff current, I_{co}.

Since I_{co} is indirectly a temperature effect, care must be taken to prevent undue temperature increases during operation. As the temperature of the transistor rises, the internal thermal agitation increases, resulting in more I_{co}. As long as the transistor is used within its ratings, however, the heat radiation rate soon equals the heat generation rate at some value of safe, stabilized cutoff current.

The collector cutoff current is generally somewhat larger for larger values of collector voltage. Manufacturers commonly provide data for reasonable magnitudes of I_{co} with several different collector voltages. If a transistor shows signs of excessive I_{co}, or if the cutoff current tends to rise continuously of its own accord, the transistor may be considered faulty.

Circuit Configurations — Common Base

Transistors and vacuum tubes are based on totally different fundamental concepts. Despite these differences, certain points of similarity do exist. This is particularly true in considering the possible circuit configurations for each of these devices.

For example, we have been discussing the common base transistor configuration in all of the material covered thus far. When we recall that the emitter of a transistor acts as the *source* of current carriers, we can at once select the cathode of the vacuum tube as being roughly equivalent. Similarly, the transistor's collector is the equivalent of the plate of the tube since both of these are the ultimate destination of the current carriers. Finally, both the transistor's base and the vacuum tube's grid serve as *control* elements, making their functions, at least, comparable in nature.

THE COMMON-BASE TRANSISTOR CONFIGURATION IS THE ROUGH EQUIVALENT OF THE GROUNDED-GRID VACUUM TUBE AMPLIFIER

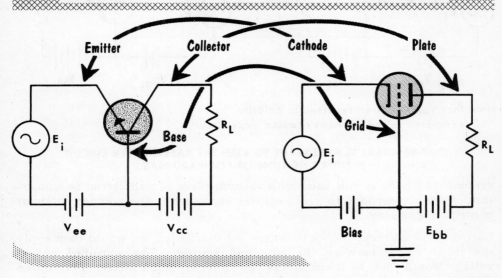

Let us set up a vacuum tube circuit in which the equivalent elements occupy similar positions. We can see from this that the common base transistor configuration, when transferred to vacuum tube circuitry, yields the familiar grounded-grid amplifier that has proved popular in f-m and television tuners.

Circuit Configurations — Common Emitter

Although the grounded-grid vacuum tube amplifier has certain applications in which it surpasses other circuits, most of us are more familiar with the conventional grounded-cathode type of amplifier in which the cathode forms the common element between grid and plate circuits. In the customary triode circuit, for example, the input signal is applied between the grid and ground (or cathode) and the output signal is taken across the plate and ground (or cathode). Making use of the apparent correspondence between transistor and tube electrodes as before, let us see what happens when the

THE COMMON-EMITTER TRANSISTOR CONFIGURATION IS SIMILAR TO THE GROUNDED-CATHODE TUBE CIRCUIT

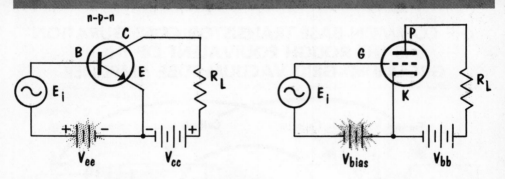

Note that V_{ee} has been reversed in polarity compared to the previous common base circuit.

THIS REVERSAL IS NECESSARY TO KEEP THE BASE-EMITTER CIRCUIT IN THE CONDITION OF FORWARD BIAS.

grounded-cathode circuit used with vacuum tubes is transferred to transistors. *Base* corresponds to *grid*; *emitter* corresponds to *cathode*; *collector* corresponds to *plate*.

Carrying the correspondence to its logical conclusion, we would then apply the input signal between the base and emitter. The amplified output voltage would then be taken across the collector and emitter. This shows that the *emitter* is the common element in this circuit just as the cathode is common to plate and grid circuits in the corresponding vacuum tube configuration. For this reason, this transistor arrangement is known as the common emitter circuit. We might mention, too, that both the common base and common emitter configurations are often called, respectively, the grounded base and grounded emitter circuits due to the fact that the common element is normally kept at signal ground potential.

Circuit Configurations — Common Emitter (contd.)

While we were studying the common base connection, we learned that the emitter-base circuit had to be biased in the *forward* direction while the collector circuit had to be biased in the *reverse* direction. Now that we

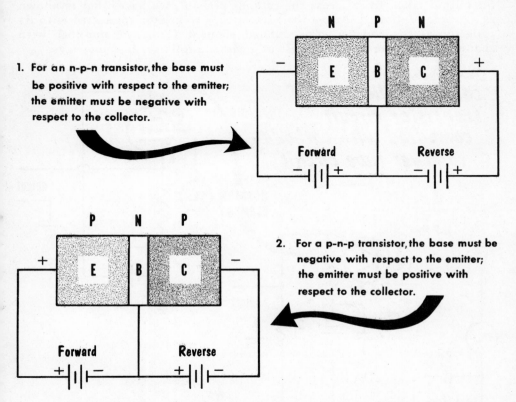

1. For an n-p-n transistor, the base must be positive with respect to the emitter; the emitter must be negative with respect to the collector.

2. For a p-n-p transistor, the base must be negative with respect to the emitter; the emitter must be positive with respect to the collector.

are beginning to alter our configurations, we should pinpoint the relative polarities of the elements somewhat more carefully for both n-p-n and p-n-p transistors to be certain we can recognize forward and reverse bias at a glance.

First, let us state a simple rule. In fundamental circuits, regardless of the configuration used, the emitter-base circuit is always forward-biased; the collector circuit is always reverse-biased. However, since the *common* element changes in each configuration, it is better to speak of bias in terms of comparative voltage polarities as follows: *for an n-p-n transistor*, the base must be positive with respect to the emitter; the emitter must be negative with respect to the collector. *For a p-n-p transistor*, the base must be negative with respect to the emitter; the emitter must be positive with respect to the collector.

Circuit Configurations — Common Collector

If we follow our present line of reasoning through to the logical end, we must determine whether or not the collector can be made the common element. In the case of the vacuum tube, when the plate is grounded and the output taken from across the cathode resistor, we have what is known as a *cathode follower*. As we shall see later, a transistor connected with its collector as the common or *grounded* element forms an amplifier with characteristics that resemble those of a cathode follower in some ways.

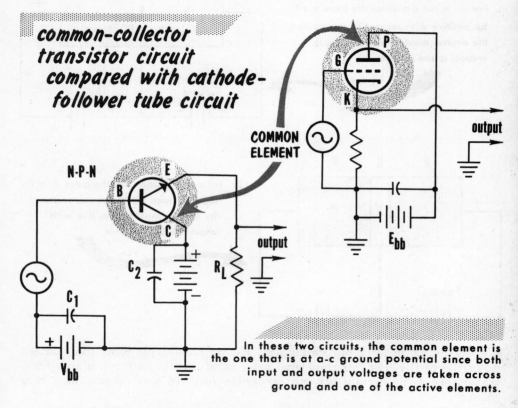

common-collector transistor circuit compared with cathode-follower tube circuit

In these two circuits, the common element is the one that is at a-c ground potential since both input and output voltages are taken across ground and one of the active elements.

These two circuits are compared in the diagrams. In making the comparison, note first that the signal input in *both* cases is applied between the ground and the control electrode (base for the transistor, grid for the tube). In the transistor circuit, capacitor C1 effectively places the bottom of the input generator at a-c ground potential. Secondly, the output in both circuits is taken between the ground and the emitting electrode. Furthermore, the collector is common to both the input and output circuits by virtue of the fact that C2 maintains it at a-c ground potential. This is an exact parallel of the plate in the case of the tube circuit.

QUESTIONS AND PROBLEMS

1. Explain why the base section of any transistor must be made as thin as possible.

2. Discuss transistor bias, giving all the details required to explain why the base-emitter circuit is always forward-biased while the collector is reverse-biased. (Include both n-p-n and p-n-p transistors.)

3. Define alpha and explain why it can never reach unity (1).

4. What is meant by resistance gain and how do we use this concept in explaining the voltage gain of a transistor in which the current gain is less than one?

5. What is meant by power gain and how is the power gain of a transistor amplifier found?

6. What is a heat sink and under what conditions is it necessary?

7. Define I_{co}. Why is this parameter so important in selecting transistors for a specific job?

8. A certain transistor is rated at 100 mw collector dissipation at 25°C and is derated 1.3 mw/C°. What would be its maximum safe dissipation if it was used in an ambient temperature of 38°C?

9. Why is alpha cutoff frequency an important consideration in the choice of a transistor for use as an i-f amplifier?

10. Discuss and compare the three transistor circuit configurations with similar triode configurations.

General Considerations for Amplifiers

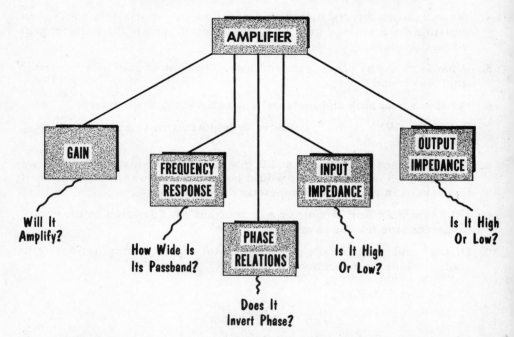

AMPLIFIER CONSIDERATIONS

These are the statistics we must have on amplifiers and what they will tell us about their performance.

AMPLIFIER

GAIN — Will It Amplify?

FREQUENCY RESPONSE — How Wide Is Its Passband?

PHASE RELATIONS — Does It Invert Phase?

INPUT IMPEDANCE — Is It High Or Low?

OUTPUT IMPEDANCE — Is It High Or Low?

Before beginning the study of transistor amplifiers, let us summarize the characteristics that are important in any amplifier circuit. *Gain*: This is the basic function of an amplifier. To be worthy of its name, it must yield an electrical gain either in the form of current, voltage or power. *Frequency response*: The frequency response characteristic of an amplifier determines whether it will be useful at high, medium or low frequencies. *Input impedance*: To get the most from an amplifier, we must always strive to arrange our input circuit so that its impedance will come close to matching that of the signal source. As it is, most input sources such as microphones and phono cartridges have tiny power capabilities. Thus, power transfer losses (even though small) may be very serious. *Output impedance*: This is a similar problem. To transfer maximum power to the load, we must know what the transistor's output impedance is so that we can match the impedance of the load to it. *Phase relations*: We are dealing with a-c signals. In many applications it is important to know whether the transistor does or does not invert the phase of the input signal.

The Common-Base Amplifier

Suppose we write all the common base characteristics in a single chart, using the figures obtained for the 2N78. What general information concerning the considerations just discussed can be gleaned from this chart? By interpreting the data correctly, we can arrive at some important conclusions. *Gain*: A power gain of 27 db, as we have shown, is a high figure. The low input and high output impedances promise a relatively high voltage gain since the alpha is very close to unity. *Frequency response*: Since the power gain is good at 0.5 mc and the output capacitance is low, we can expect good frequency response between 0 and 500 kc. (Note also the high alpha cutoff frequency.) *Input impedance*: Very low as indicated by the value of 55 ohms. This would call for a low-impedance source. *Output impedance*: Very high. This, of course, requires a high-impedance load for good matching. *Phase relations*: The phase relationship between input and output cannot be determined from the characteristics. We will want to analyze the action of the circuit to determine what this is.

2N78 n-p-n CHARACTERISTICS

ABSOLUTE MAXIMUM RATINGS (25°C)

Collector-to-emitter voltage (base open) V_{ce}	15 volts
Collector-to-base voltage (emitter open) V_{cb}	15 volts
Collector current (I_c)	20 ma
Emitter current (I_e)	-20 ma
Maximum collector dissipation (P_c)	65 mw
Maximum junction temperature (T_J)	100°C
Maximum storage temperature (T_{stg})	85°C

SIGNAL CHARACTERISTICS

Current gain (common-base) (α)	.983
Alpha cutoff frequency (f_{ab})	9 mc
Input impedance (Z_i) for 270 cps, output short-circuited	55 ohms
Output impedance (Z_o) for 270 cps, input open-circuited	5 megohms
Noise Figure (NF) with V_{cb} = 1.5 v, I_e = -0.5 ma, f = 1 kc	12 db
Output capacitance (C_o), f = 2 mc	0.3 $\mu\mu$f
Power gain (PG) at 500 kc with V_{cb} = 1.5 v, and I_e = 1.0 ma	27db
Collector cutoff current (I_{co}), V_{cb} = 15v	5 μa
V_{cb} = 5v	0.7 μa

Review of Vacuum Tube Phase Relations

In the vacuum tube circuit, the grid is negative with respect to the cathode due to the voltage drop in R_k before a signal is applied. At the same time, the potential of the upper output terminal is less positive (more negative) than the positive terminal of the plate source V_{bb} because a steady plate current flows through R_L. This causes a voltage drop across this resistor, having the polarity shown in the illustration.

Now let us apply a postive-going half-cycle of signal voltage. This drives the grid more positive than it was previously, causing the plate current of the tube to increase. Thus, the voltage drop across R_L increases, causing a further reduction of the voltage at the upper terminal. Hence, a *positive*-going input half-cycle causes a *negative*-going output half-cycle as shown, resulting in a 180° phase shift between input and output. We call this *phase inversion*. A similar phase reversal occurs when the input half-cycle is

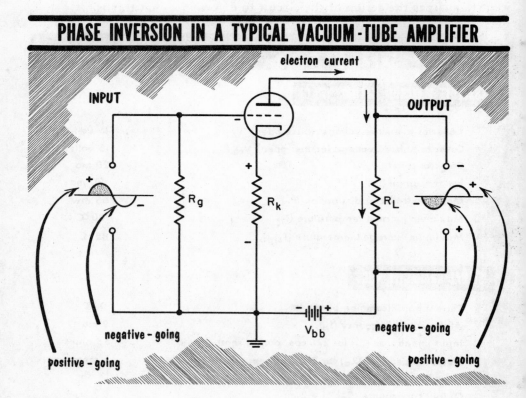

PHASE INVERSION IN A TYPICAL VACUUM-TUBE AMPLIFIER

negative-going. The grid is driven negative, the plate current decreases, the drop across R_L decreases and the upper output terminal becomes more *positive*.

Phase Relations in a Common-Base Amplifier

The base-emitter circuit is forward-biased. In the case of the p-n-p transistor used as an example, this means that the emitter is more positive than the base. When a positive-going half-cycle is now inserted in series with V_{ee},

THERE IS NO PHASE INVERSION IN A COMMON-BASE TRANSISTOR AMPLIFIER

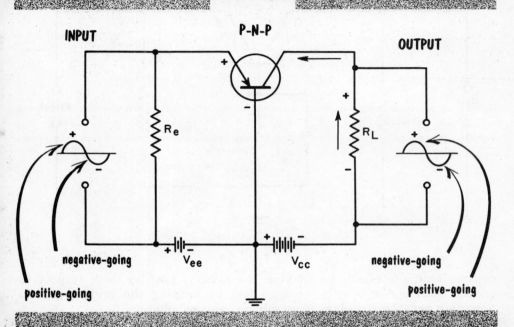

the emitter becomes more positive than before, increasing the emitter-base current. In a transistor, an increase of emitter-base current produces a corresponding increase of collector current. Since the direction of current flow is upward in R_L, the top terminal of this resistor must become more positive than it was before with respect to the bottom terminal. Hence, a *positive*-going input half-cycle gives rise to a *positive*-going output half-cycle. This means that there is *no phase inversion* in the common base transistor amplifier.

Phase relations are important considerations in many types of multistage amplifiers, in oscillators, and in video amplifiers for television. It is important to remember how we determine whether phase inversion does or does not occur for future use.

Single-Battery Common-Base Circuit

A vacuum tube amplifier circuit is normally designed to do away with the need for a grid-bias battery by utilizing the voltage drop across a cathode

The Common-Base Circuit Modified for Single-Battery Operation

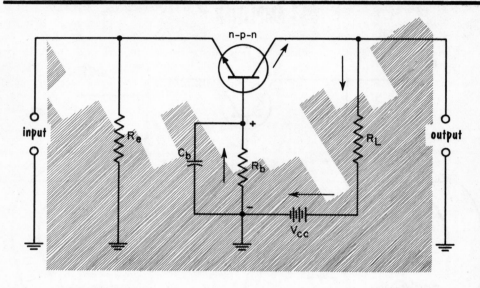

resistor (cathode bias). By making the cathode positive with respect to ground (common reference voltage) and by keeping the grid at ground potential, negative grid-bias is established as needed.

The same procedure may be followed to eliminate emitter battery V_{ee} in the common-base circuit. We need merely add a base resistor (R_b) and make the lower terminal of this resistor common to both input and output circuits. In addition to doing away with the emitter battery, this circuit makes it possible to maintain one input and one output terminal at ground potential. This makes the transistor circuit appear much like the conventional vacuum tube amplifier since there is now a common reference potential (ground) for both input and output. The small I_{co} flowing through R_b places the base at a slightly higher positive potential than ground. Since the emitter is connected to ground through R_e, this element must be negative with respect to the base (this is an n-p-n transistor). Thus, the small amount of forward-bias is obtained without the need for a battery. As in vacuum tube amplifiers, the bias resistor R_b may be bypassed to maintain the voltage drop across it constant, despite the possible presence of alternating signal currents.

A Practical Common-Base Amplifier

Note that transformer coupling is used in both input and output circuits. This arrangement permits us to see how important it is to know the input and output impedances of the transistor. The input and output impedances of the 2N265 audio transistor, as given by the manufacturer for the common-base circuit are, respectively, 29 ohms and 2 megohms. These figures account for the transformers used: the input transformer has a high-impedance primary to match a microphone or a phono-cartridge, while its secondary impedance is low to match the emitter-base impedance. Similarly, the primary of the output transformer matches the high output impedance of the transistor while its secondary impedance is governed by the type of load (in this case a 1000-ohm headset). In addition, the typical collector supply voltage as given in the handbook is 12 volts, easily obtained by two 6-volt dry batteries or mercury batteries in series. The battery polarization informs us that the 2N265 is a p-n-p transistor, since the collector is negative with respect to common ground.

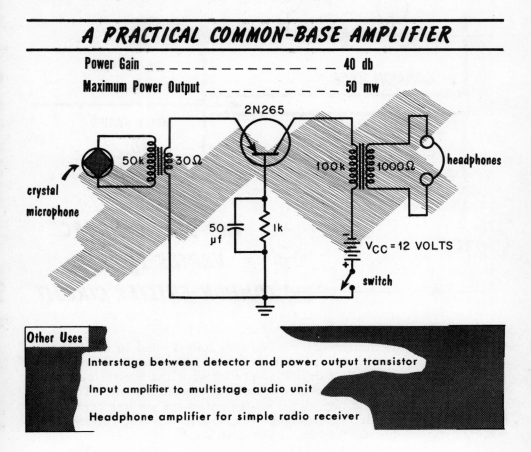

A PRACTICAL COMMON-BASE AMPLIFIER

Power Gain _ _ _ _ _ _ _ _ _ _ _ _ _ _ 40 db

Maximum Power Output _ _ _ _ _ _ _ _ _ _ 50 mw

2N265

50k   30Ω   100k   1000Ω   headphones

crystal microphone

50 µf   1k

V_{CC} = 12 VOLTS

switch

Other Uses

Interstage between detector and power output transistor

Input amplifier to multistage audio unit

Headphone amplifier for simple radio receiver

The Common-Emitter Circuit

We have drawn the basic circuits for the common-base and common-emitter configurations to compare them in certain essential respects. Both transistors shown are n-p-n. We could replace them by p-n-p types merely by reversing the polarities of both batteries. Here are the factors to compare: (1) Even though the configurations are completely different in connection, *the same battery polarities are used in both cases.* For the n-p-n types, the emitter is negative with respect to the collector while the base is positive

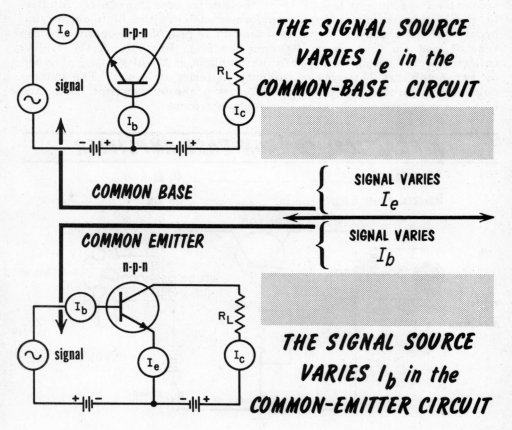

THE SIGNAL SOURCE VARIES I_e in the COMMON-BASE CIRCUIT

SIGNAL VARIES I_e

SIGNAL VARIES I_b

THE SIGNAL SOURCE VARIES I_b in the COMMON-EMITTER CIRCUIT

with respect to the emitter. This establishes forward-bias in the base circuit and reverse-bias in the collector circuit. (2) The base current is very small in both circuits. It is the difference between the emitter current I_e and the collector current I_c regardless of configuration used. (3) In the common-base connection, the signal source varies the *emitter current.* In the common-emitter circuit, the signal source varies the *base current.* This is a very important consideration, as you shall see.

The Meaning of Beta

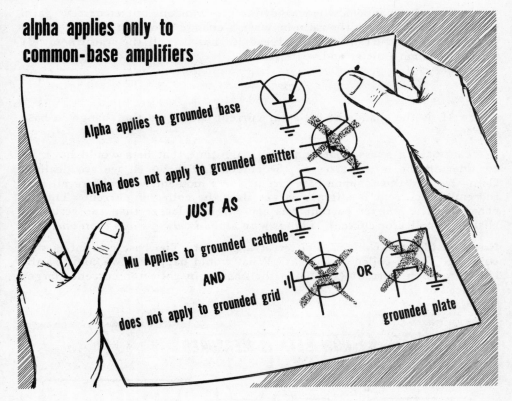

alpha applies only to common-base amplifiers

Alpha applies to grounded base

Alpha does not apply to grounded emitter

JUST AS

Mu Applies to grounded cathode

AND

does not apply to grounded grid OR grounded plate

In the common-base circuit it was shown that current amplification is a ratio set up between the change of collector current and the change in the causative emitter current. That is:

$$\alpha = \frac{\Delta I_c}{\Delta I_e}$$

We also saw that alpha can never exceed (or even attain) unity because the collector current is always smaller than the emitter current, with the base current representing the difference between them.

Now, however, we are about to vary the *base current* to observe the effect upon the collector current. Since alpha has already been defined as the ratio of the change of collector current to the change of emitter current, it could not possibly retain its meaning if the base current is varied instead. Thus, alpha has a definite and specific connection with the common-base configuration. It cannot be used to measure current amplification in the common-emitter circuit. We need, therefore, a new constant that will apply specifically to the current gain of the common-emitter configuration.

The Meaning of Beta (contd.)

Let us see what happens when we define the common-emitter current gain. Current gain is merely the ratio between a change in output current to the change in input current that produced it. Let us call this ratio *beta* (β) for the common-emitter and write it thus:

$$\beta = \frac{\Delta I_c}{\Delta I_b}$$

where ΔI_c is the change of collector current caused by the change in base current ΔI_b.

We can make a guess relative to the magnitude that beta would have, by considering the initial sizes of the currents with which we are dealing. Going back to the common-base circuit for a moment, you will recall that the emitter current is slightly larger than the collector current. Thus, a change in the emitter current was also slightly larger than the resulting change in collector current. *This is why alpha is always less than one.*

Now consider the common-emitter arrangement. The base current is tiny compared to the collector current. What would happen if we compare the change of base current (small) to the change in collector current (large) produced by it?

HOW BETA IS MEASURED

METHOD:

1. With R set on position A, I_b and I_c are both recorded. Call these readings I_{bA} and I_{cA}

2. Change wiper of R to position B. Again read base and collector currents. Call the new readings I_{bB} and I_{cB}.

3. Beta is now found from the ratio $I_{cA} - I_{cB} \big/ I_{bA} - I_{bB}$.

Note: We have been referring to $I_{cA} - I_{cB}$ as ΔI_c, and $I_{bA} - I_{bB}$ as ΔI_b

The Meaning of Beta (contd.)

When the experiment suggested by the preceding drawing is performed, it is found that a small change in base current caused by shifting the position of the wiper on the input potentiometer, causes a relatively *large change in the collector current.* That is, the ratio $\Delta I_c/\Delta I_b$ or beta turns out to be quite large.

For example, the beta rating for an average 2N78 (the transistor whose ratings we examined in detail) is given by the manufacturer as 24. This means that a change of 1 microampere of current in the base circuit will result in a change of 24 microamperes in the collector circuit.

ALPHA AND BETA COMPARED FOR TWO TYPICAL TRANSISTORS

	2N78	2N265	
BASE CURRENT GAIN β	28	110	$\beta = \dfrac{\Delta I_c}{\Delta I_b}$
CURRENT GAIN α	0.966	0.991	$\alpha = \dfrac{\Delta I_c}{\Delta I_e}$

As another example, consider the 2N265 which we have also mentioned. This transistor is intended for audio driver service. It is a typical audio amplifier used in transistor radios, following the detector stage, to provide sufficient gain to drive the output transistor through a coupling transformer. When the collector voltage is −5 volts (the 2N265 is a p-n-p type, hence the negative sign), and the emitter current is 1.0 ma, the beta is given as 110. Thus, the common-emitter connection is capable of substantial current gain as contrasted with common-base.

Common-Emitter Characteristics — Collector Curves

Just as we can study the performance of tubes from an analysis of the plate characteristic curves, we can learn much about the behavior of transistors from the *collector characteristic curves.*

The procedure for obtaining the curves is evident from their structure. I_{ceo}, the collector cutoff current, with the base open-circuited, is determined by leaving the base circuit open while potentials from 0 to 15 volts are

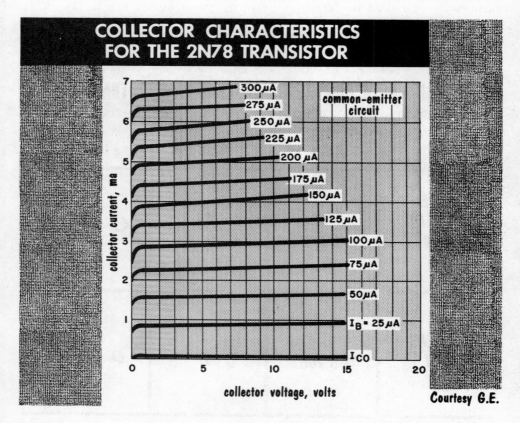

COLLECTOR CHARACTERISTICS FOR THE 2N78 TRANSISTOR

Courtesy G.E.

applied in succession to the collector. The current for each collector voltage is then recorded as points which, when connected, form the I_{ceo} line. A controllable voltage source is then connected across base and emitter and the base current adjusted to 25 microamperes. Again, the collector voltage is varied to provide the coordinate for the $I_b = 25 \mu a$ line.

This process is repeated in steps of 25 μa all the way up to 300, while the collector current corresponding to the various collector voltages is read off and used to construct the family of curves as it appears in the chart.

Using the Collector Curves

One of the first interesting things that strikes you as you look at the collector curves is that the current does not rise very rapidly with changes of collector voltage. Notice how horizontal the graph lines are, especially when the base current is low. Even for higher base currents, the slopes are very shallow. You may remember that the transistor characteristics, in this respect at least, resemble the plate characteristics of pentode vacuum

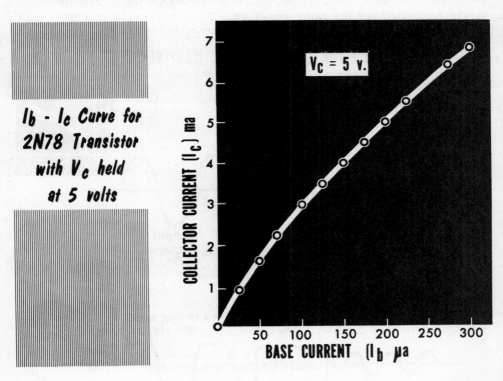

I_b - I_c Curve for
2N78 Transistor
with V_c held
at 5 volts

tubes. We can say it this way: *over the range recommended by the manufacturer, the collector current of a transistor is relatively independent of the collector voltage.*

On the other hand, a small change of base current always produces a relatively large change in collector current, if the collector potential is maintained constant. Does the collector current rise in equal steps for equal increments of base current? To see this, choose a certain collector potential, say 5 volts, then follow the 5-volt line upward and note how the current changes in each step of 25 μa base current. To help you evaluate the collector current changes, we have drawn I_b–I_c curve for a collector potential of 5 volts.

Using the Collector Curves (contd.)

Although this curve is not a perfect straight line, it certainly does approach it closely. If the transistor is used in an audio amplifier circuit, for example, we might then say: *If the range of variation of I_b is held within reasonably small limits, the collector current varies linearly with it.*

This may be interpreted as follows: if the input current to the base is a weak audio current, the current variations in the collector circuit will be larger *but will have the same waveform.* Thus, we arrive at a measure of

A CURVED TRANSISTOR CHARACTERISTIC MAY INTRODUCE DISTORTION

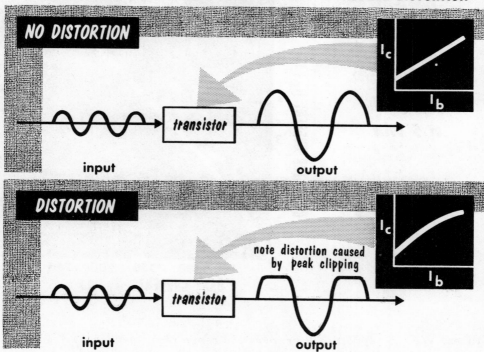

the *fidelity* or, conversely, the inherent distortion that we may expect from a transistor of this type. As long as the I_b–I_c curve is a straight line, the inherent distortion will be zero. The greater the curvature of the graph, the greater will be the distortion that can be attributed to the transistor itself.

All of the foregoing presupposes that the circuit components have been selected to produce the proper bias for the specific transistor used. If this is not the case, distortion may occur that has no connection with the inherent characteristics of the transistor.

Using the Collector Curves (contd.)

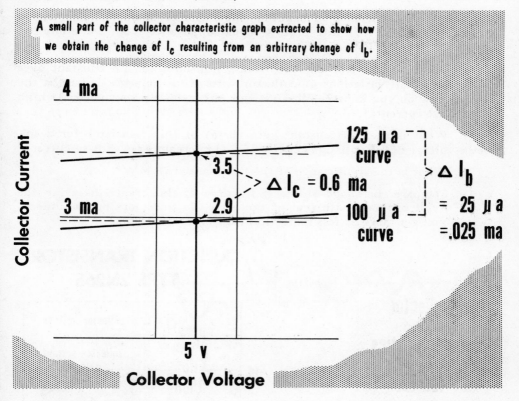

A small part of the collector characteristic graph extracted to show how we obtain the change of I_c resulting from an arbitrary change of I_b.

Either the collector characteristic curves or the I_b–I_c curve may be used to estimate the beta of a transistor in just a few seconds. Since the collector characteristics are those generally found in manufacturers' rating sheets, suppose we use these to check the beta, or *base current gain* as it is often called, of the 2N78.

Since this transistor normally operates at a collector potential of 5 volts, we first locate the 5-volt line. Then we select an average region of base current change along this line, say from 100 microamperes to 125 microamperes. Using our knowledge that beta = $\Delta I_c / \Delta I_b$, we can take the change of I_b from 100 μa to 125 μa for a total of $\Delta I_b = 25$ μa. We then note that I_c goes from 2.9 ma to 3.5 ma over this range of I_b. Substituting these values in the equation we have:

$$\beta = \frac{3.5 - 2.9}{.025} = \frac{0.6}{.025} = 24 \quad (25 \ \mu a = 0.025 \ \text{ma})$$

Note that this checks nicely with the figure given some pages back for the beta of a 2N78.

75

Practicing With the Collector Characteristics

To gain some experience in handling the collector characteristics and to increase your facility in interpreting them, suppose we pause and study the curves for the 2N265 audio driver transistor.

(1)   Does the collector current of this transistor remain constant over reasonably small variations of collector voltage to the same degree as the 2N78? How do you know? Where is this effect more pronounced, At high or low base currents?

(2)   What is the base current gain (beta) of this transistor for a collector voltage of −5 volts between 22 and 33 microamperes of base current?

(3)   What is the significance of the line marked 75 MW?

(4)   Compare the shape of the I_c–I_b curve of this transistor along the $I_c = -5$ volt ordinate with that of the 2N78. Is this more or less linear? Would you get more or less distortion using this transistor?

collector characteristics for the 2N265 transistor

Courtesy G.E.

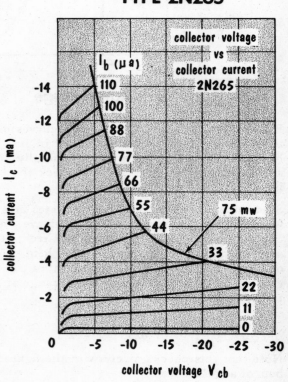

JUNCTION TRANSISTOR TYPE 2N265

Answers to the Questions on P. 76

TRANSISTOR POWER DISSIPATION IN THE COLLECTOR CIRCUIT CAN BE OBTAINED FROM THE COLLECTOR CHARACTERISTICS.

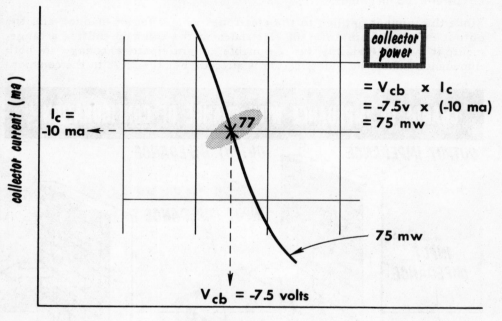

collector power

$= V_{cb} \times I_c$
$= -7.5v \times (-10\ ma)$
$= 75\ mw$

$I_c = -10\ ma$

75 mw

$V_{cb} = -7.5$ volts

collector current (ma)

collector voltage

(1) No. The curves have a noticeable upward slope, becoming increasingly pronounced in the region of the higher base currents. Thus, in this transistor, the collector current is not quite as independent of the collector voltage as in the previous case. (2) Between 22 and 33 microamperes of base current, the collector current changes from approximately −2 ma to −3.2 ma for a total increase of 1.2 ma or 1200 microamperes. Thus, the beta is: 1200/11 = 109. This is sufficiently close, considering approximations in reading the co-ordinates of the curve, to the previous value as given by the manufacturer, $\beta = 110$. (3) The line marked "75 mw" is the maximum power dissipation line for this transistor. If we select any point on this line, and read the current and voltage as the co-ordinates of this point, we will find that their product is approximately 75 milliwatts. (We have selected a point at random in our drawing to show this.) (4) The linearity of the 2N265 is not as good as that of the 2N78. Hence, the 2N265 will produce more distortion. This increase in distortion has come about as a result of the large increase in current gain. It would not be great enough, however, to be disturbing.

Impedances in the Common-Emitter

To refresh our memories relative to the input and output impedances of the common-base circuit for the 2N78, we write again: Input impedance (Z_i), output short-circuited . . . 55 ohms; output impedance (Z_o), input open-circuited . . . 5 megohms.

Since the input is applied to the electrodes in a different manner, and the output is taken from different electrodes in the common-emitter arrangement, it would certainly be reasonable to anticipate a change in both impedances when the configuration is altered. For the 2N78 in the common-

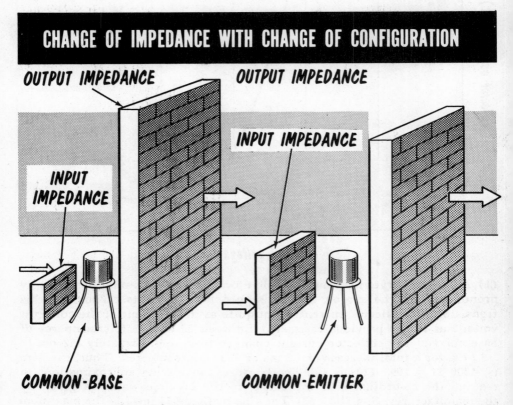

CHANGE OF IMPEDANCE WITH CHANGE OF CONFIGURATION

OUTPUT IMPEDANCE   OUTPUT IMPEDANCE

INPUT IMPEDANCE

INPUT IMPEDANCE

COMMON-BASE   COMMON-EMITTER

emitter circuit, the following impedances are specified in the characteristics table: Input impedance (Z_i), output short-circuited . . . 350 ohms; output impedance (Z_o), input short-circuited . . . 15,000 ohms.

Note that the input impedance is still relatively low, although it is higher than before; the output impedance, however, drops considerably (from 5,000,000 ohms to 15,000 ohms). The first significance of these changes is, of course, that some additional attention will have to be paid to the problem of input and output impedance matching.

Voltage and Power Gain in the Common-Emitter

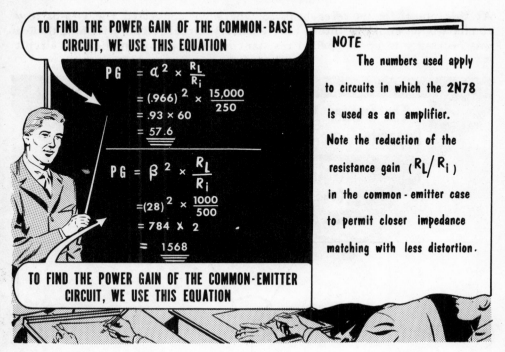

TO FIND THE POWER GAIN OF THE COMMON-BASE CIRCUIT, WE USE THIS EQUATION

$$PG = \alpha^2 \times \frac{R_L}{R_i}$$

$$= (.966)^2 \times \frac{15,000}{250}$$

$$= .93 \times 60$$

$$= 57.6$$

$$PG = \beta^2 \times \frac{R_L}{R_i}$$

$$= (28)^2 \times \frac{1000}{500}$$

$$= 784 \times 2$$

$$= 1568$$

TO FIND THE POWER GAIN OF THE COMMON-EMITTER CIRCUIT, WE USE THIS EQUATION

NOTE
The numbers used apply to circuits in which the 2N78 is used as an amplifier. Note the reduction of the resistance gain (R_L/R_i) in the common-emitter case to permit closer impedance matching with less distortion.

In working circuits, it may be demonstrated that the voltage gain of the common-base configuration is somewhat higher than that of the common-emitter arrangement, although not as much higher as we might expect. Although the current gain, alpha, of the common-base is less than that of the common-emitter circuit, the change in impedances more than makes up for it. Put simply, the increase in current gain in the common-emitter is more than offset by the loss of resistance gain.

In the realm of *power gain*, however, the common-emitter is definitely superior. As we look at the sample solutions for the 2N78 in the illustration, we can see at once that the current gain increase when we go from common-base to common-emitter becomes very important. Since power gain depends upon the *current gain squared*, as we go from an alpha of 0.966 to a beta of 28, the power tends to go up as the square of 28 or 784 times due to this alone. Then, when the beta squared is multiplied by the resistance gain, we arrive at a power gain of 1568. Notice also that the ratio of R_L to R_i has been made much smaller in the common-emitter case. This is due to the change of impedances when common-base is changed to common-emitter as discussed earlier. Although the resistance gain has decreased in the common-emitter, the power gain has increased due to the increase in current gain.

The Single-Battery Common-Emitter Circuit

At this point, let us refer back to where we converted the two-battery common-base amplifier into a single-battery circuit. You will find that it was necessary to add a special resistance and capacitor in the base circuit

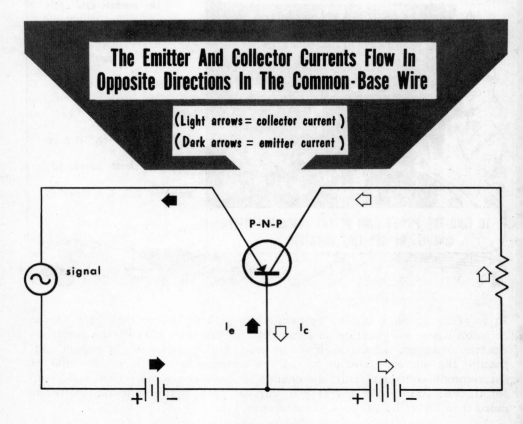

The Emitter And Collector Currents Flow In Opposite Directions In The Common-Base Wire

(Light arrows = collector current)
(Dark arrows = emitter current)

to provide the required biasing potentials. That is, R_b permits the base to have the correct polarity with respect to the emitter (forward bias) and with respect to the collector (reverse bias).

If we analyze the two-battery common-base circuit further, it is apparent that the reason for the need of an extra biasing component lies in the directions of flow of the emitter and collector currents. Using a p-n-p transistor as an example (the n-p-n is analyzed the same way except that all current directions are reversed), we see that I_e and I_c flow in *opposite* directions in the common lead going to the base. Since one battery, placed anywhere at all in the circuit, could never produce two oppositely-directed currents in a common element, it is necessary to create the correct bias potentials by means of an *artificial* aid — the bias resistor.

The Single-Battery Common-Emitter Circuit (contd.)

Now let us analyze the two-battery common-emitter circuit. V_{ee} forces current up through R_i (or through the signal source itself if the source has d-c continuity), through the base to the emitter, then downward through the common lead back to the positive terminal of the battery as shown by the dark arrows. V_{cc} drives current up through R_L, through the collector to the emitter, and then downward through the common connection back to the positive end of this battery as shown by the light arrows.

Thus, the currents in the common wire from the emitter to the junction of the two-battery positives, flow in the same directions. This should make it easy to install a battery so placed that it does the jobs of both the original

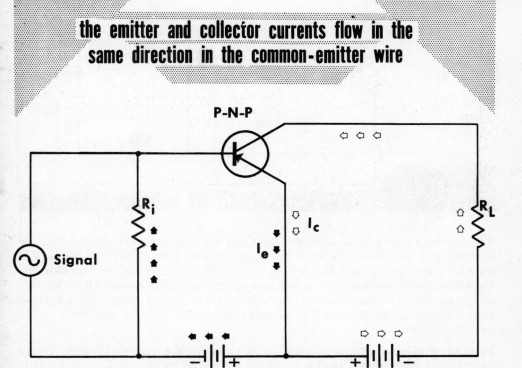

the emitter and collector currents flow in the same direction in the common-emitter wire

P-N-P

R_i

Signal

I_c

I_e

R_L

V_{ee}

V_{cc}

sources. Wherever we place it, remember that the base must be more negative than the emitter but not as negative as the collector for the p-n-p type.

The Single-Battery Common-Emitter Circuit (contd.)

A Simple One-Battery Common-Emitter Amplifier Circuit

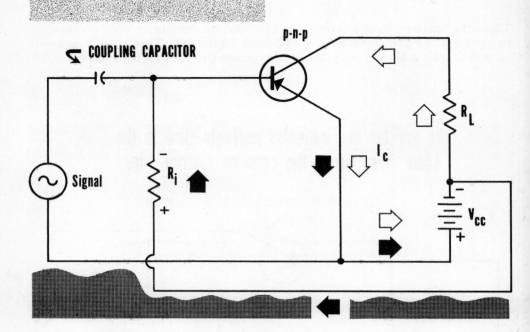

As a matter of fact, it is so easy to do away with the second battery that we wonder why two batteries were ever used in common-emitter circuits. We merely note that V_{ee} and V_{cc} are both negative with respect to the emitter. Checking over the connection, we can recognize immediately that the collector current follows exactly the same path as before: from the negative end of the battery, through the collector, out of the emitter and back to the positive terminal.

Simultaneously, the same battery sends some current, small though it may be, into the base circuit in the direction indicated by the dark arrows. This current flows in exactly the same manner as when two batteries were used. All that remains is to select a suitable value for R_i so that the base voltage has the right magnitude. Since the size of R_i depends, to a great extent, upon the type of transistor, the potential of V_{cc} and the ambient temperature, we cannot state a specific value for it. A typical value of R_i for a 2N78 at room temperature in a standard i-f circuit is about 10,000 ohms.

Why More Thought Must be Given to Biasing

When you look at the common-emitter circuit just presented, and then glance through some practical amplifiers being used today, you may be struck by the fact that none of them are quite as simple as the one we have just drawn. Why should this be?

When the base-emitter circuit is forward-biased by the small voltage drop across R_i, a definite base current will flow. Due to the current amplification (beta) of this transistor configuration, the base current will cause a very definite collector current which will be substantially higher than the cutoff current, I_{co}. Thus, the zero-signal current in the collector circuit will depend to a great degree upon the beta of the particular transistor.

Unfortunately, the betas of even identical types varies to a certain extent. It is always good procedure to make the operating conditions as indepen-

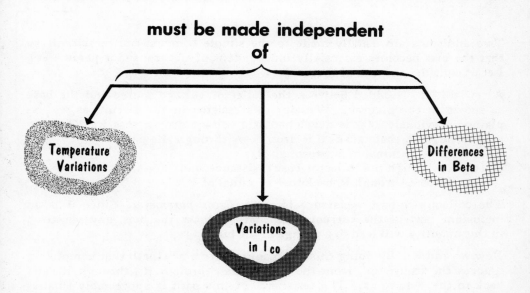

dent as possible of the transistor's *parameters*. (A parameter is any controlling factor in an electronic circuit which may or may not be held constant. Beta is such a parameter).

Practical biasing is usually accomplished by some sort of simple feedback network that helps make the circuit operation independent of the parameters.

Practical Biasing

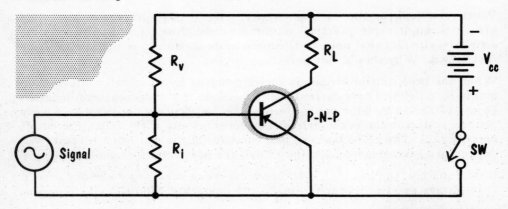

ADDING BIAS RESISTANCE TO THE COMMON-EMITTER CIRCUIT

Two additions are usually made to the simple common-emitter circuit so that the bias becomes essentially independent of the transistor parameters. Let us consider the first of these.

A resistor, R_v, is added between the collector voltage source and the base as shown in the diagram. Without this resistor in the circuit, as we explained previously, bias is developed as a voltage drop across R_i. The current through R_i that causes this drop flows through the path:

from the — terminal of V_{cc} through R_L;
from R_L through the collector-base resistance, and
from the base through R_i back to + terminal of V_{cc}.

The collector-to-base resistance is a *transistor parameter*. Since it is an important part of the current path that produces the bias, any variation in this quantity will lead to a change in the bias.

Now we add R_v. By doing this, we manufacture a new path that completely ignores the transistor: from the — terminal, through R_v, through R_i, and back to the + terminal. The resistance of this path is appreciably smaller than the original one, hence it plays a much more important role in determining the voltage drop across R_i than the smaller current does.

R_i must be changed in value when R_v is added to assure the same value of bias originally decided on. With both resistors in the circuit, the drop across R_i is then much more dependent upon the fixed resistors than it is on the transistor's behavior.

Practical Biasing (contd.)

The second addition to the bias circuit is the resistor R_e. This resistor is added to stabilize the performance of the transistor, particularly with regard to temperature effects. It produces a degenerative effect just as does the cathode resistor in a vacuum tube amplifier, and for this reason is often called a *feedback* resistor.

We have seen that I_{co} is a temperature effect. As the temperature of the transistor rises, the internal thermal agitation increases, resulting in more I_{co}. This changes the operating conditions of the transistor and would tend to affect the bias. Furthermore, the effect may become cumulative in the following way: suppose the temperature of the transistor rises slightly due to an increase in the ambient temperature; I_{co} would then increase causing further power to be dissipated in the collector junction, again raising the temperature and, indirectly, the size of I_{co}. This process can go on until the transistor destroys itself unless R_e is added to the circuit.

Now any increase in I_{co} will develop a bucking voltage across R_e which reduces the effective forward bias of the base-emitter circuit by making the emitter more negative than it was previously. A lessening of *forward* bias causes the base current to go down. This is reflected in the collector circuit as an effect that causes the collector current to decrease. Remember that a decrease of base current causes a decrease of collector current due to normal transistor action. Hence, the original rise of I_{co} is *cancelled* by the action of R_e to bring about vastly improved bias stability.

In practical amplifier circuits, as in vacuum tube designs, the resistor R_e is bypassed by a large capacitor to prevent degeneration of the signal. What we want here is a d-c action but we do not want the signal to be affected.

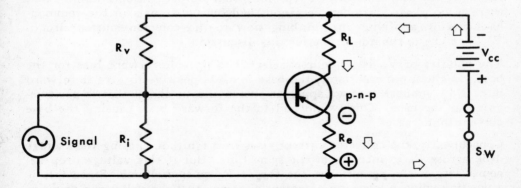

the addition of R_e stabilizes transistor performance

Phase Relations in the Common-Emitter Circuit

THE OUTPUT AND INPUT IN A COMMON-EMITTER AMPLIFIER ARE 180° OUT-OF-PHASE

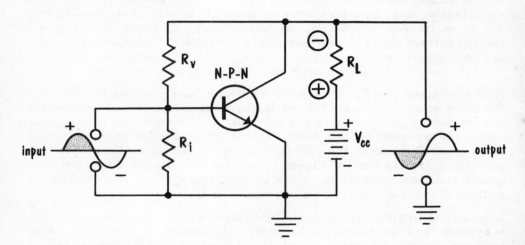

Let us follow the same line of reasoning that we used in determining that there is no phase shift between input and output signals in the common-base circuit, applying our thinking now to the common-emitter circuit. We'll return to the n-p-n type for this discussion.

The polarity of V_{cc} is, of course, selected to provide forward bias for the base. In the n-p-n transistor, the base is made positive to give it forward bias. Now suppose that we apply an instantaneous signal that is positive-going at the base. This would add to the forward bias, causing the base current to increase.

Consequently, the collector current rises in magnitude causing the voltage drop across R_L to increase at the same time. But if this voltage drop becomes larger, the *potential at the top of R_L must diminish*. Since this is a positive point, a lowering of potential means that the voltage is going in a negative *direction*. Essentially, then, feeding a positive-going pulse to the input of the system results in a negative-going pulse at the output. Hence, in the common-emitter circuit, the signal input is always out of phase with the output by 180°.

The Common-Collector Circuit

Suppose you try a little exercise in "transistor thinking" at this point. Before referring to the diagram, try to answer these questions by using facts already learned.

(1)   We have studied the basic hookups for the common-base and common-emitter circuits. Obviously, if we follow the pattern, we should be able to draw a common-collector circuit. If so, will the base still be forward biased? Will the collector be reverse biased?

(2)   Depending upon your answer to (1), and assuming that we want to use an n-p-n transistor, should the base be positive or negative with respect to the emitter?

(3)   Should the emitter be positive or negative with respect to the collector?

(4)   Answer the same questions for a p-n-p type.

(5)   Draw the d-c bias diagram for a common-collector n-p-n transistor.

(6)   Draw the d-c bias diagram for a common-collector p-n-p transistor.

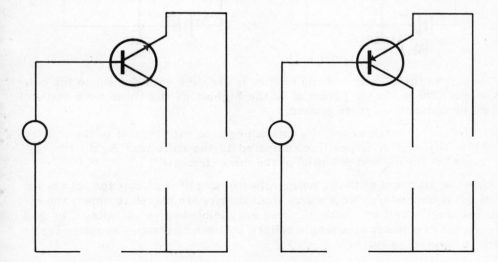

Complete the N-P-N and P-N-P Common-Collector Circuits

The Common-Collector Circuit (contd.)

If you answered the preceding questions correctly, you should have arrived at the following conclusions:

(1)   Regardless of the configuration, the base is always forward biased and the collector is reverse biased.

(2)   and   (3)   Here again, configuration type does not affect the bias polarities in the least. In an n-p-n transistor the base is always positive with

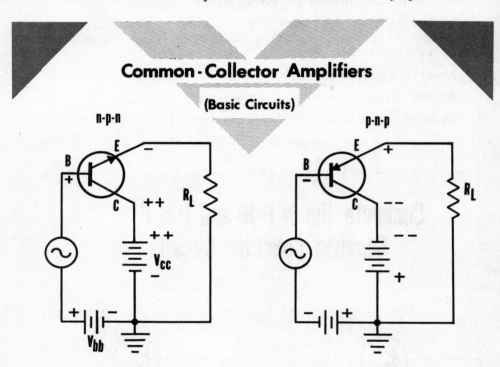

Common - Collector Amplifiers

(Basic Circuits)

respect to the emitter and the emitter is negative with respect to the collector. The collector potential is the highest of the three with respect to the common point, or ground.

(4)   In the p-n-p transistor, the base is negative with respect to the emitter while the emitter is positive compared to the collector. Again the collector has the highest potential of the three elements.

Study the diagrams of the common-collector amplifier circuit for both n-p-n and p-n-p transistors. Note where the batteries are placed to obtain the required bias polarities. Both of these are double-source circuits. Can you figure out how to set up a single battery in place of the two shown? Try it before proceeding further.

Common-Collector Single-Source Circuit

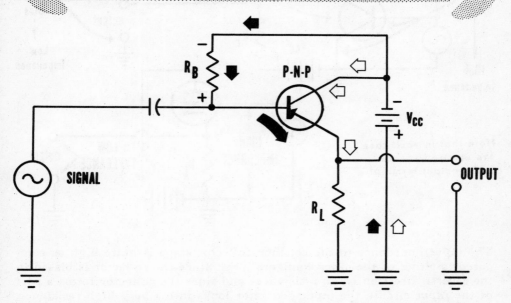

single-battery common-collector amplifier

We are always interested in reducing the number of batteries required to provide bias for transistor circuits. In the case of the common-collector amplifier, we want to use one battery to obtain the proper voltage between emitter and collector for static biasing, and if possible, a resistor correctly placed to establish the desired base bias. This is easily accomplished by connecting a resistor of the right size directly from V_{cc} to the base. Note that the base battery has been omitted.

Analyzing this circuit, using a p-n-p transistor as an example, we start by tracing the current from the negative terminal of the source. We have shown this current as both dark and light arrows, since part of the total battery current will go one way and part another way before we are through. Most of the battery current flows through the collector circuit back to + terminal of the battery; this is the collector current (light arrows). A small portion of it, however, goes through R_b into the base. It crosses the emitter-base junction, goes to ground and then to the + terminal of V_{cc}. There is then a voltage drop across R_b (polarity shown) that makes the base slightly more negative than the emitter — the condition necessary for proper bias on a p-n-p transistor.

Impedances in Common-Collector Circuit

common-collector circuit compared with cathode-follower

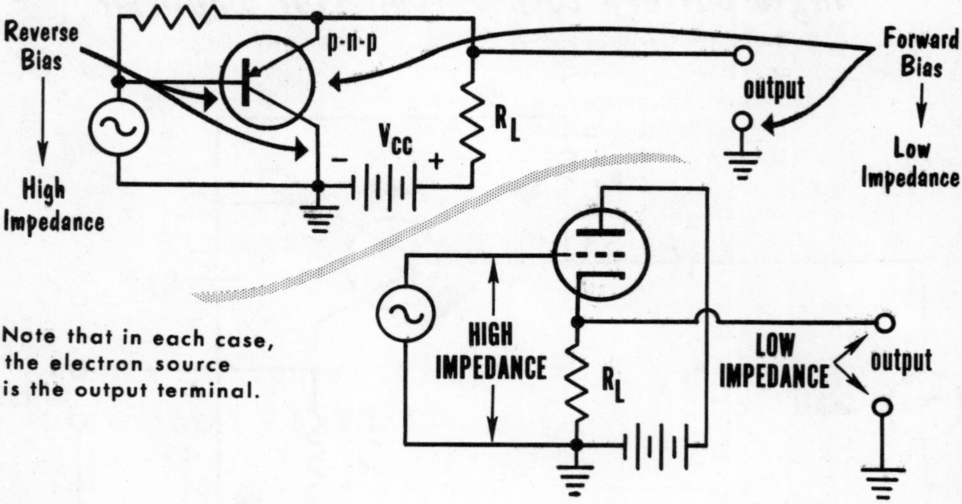

Reverse Bias → High Impedance

p-n-p

R_L

V_{CC}

output

Forward Bias → Low Impedance

Note that in each case, the electron source is the output terminal.

HIGH IMPEDANCE

R_L

LOW IMPEDANCE

output

The input impedance to the common-collector stage is quite high as compared to either of the other configurations. Since the collector is biased in the reverse direction (high resistance) and since the collector forms a part of the *input* circuit, the input generator looks into a very high resistance. On the other hand, the emitter is biased in the forward (low resistance) direction and forms a part of the output circuit. Thus, we would anticipate a *low output* impedance. This analysis turns out to be correct. Compare the input and output impedances of the common-collector stage with the other two configurations by examining the chart below:

IMPEDANCES (AVERAGE) FOR A 2N78

	Common-base	Common-emitter	Common-collector
Input	55 ohms	350 ohms	650,000 ohms
Output	5,000,000 ohms	15,000 ohms	150 ohms

It is interesting to note, too, that the common-collector configuration has characteristics similar to those of the vacuum tube cathode follower. Both have high input and low output impedances as shown in the comparative drawings. In each, the element that is usually considered the source of electron flow — the cathode in the tube and the emitter in the transistor — forms a part of the output circuit.

Common-Collector Voltage and Power Gain

The comparison of a common-collector transistor amplifier and a vacuum tube cathode follower may be carried even further. The voltage gain of a cathode follower can approach, *but never attain*, a value of 1. The same is true of the common-collector amplifier. In practical circuits the voltage gain may range anywhere from 0.7 to 0.9, never higher.

Furthermore, measurements indicate that the power gain of a typical common-collector stage is considerably lower than either the common-base or common-emitter configuration. We have shown that the 2N78, can be ex-

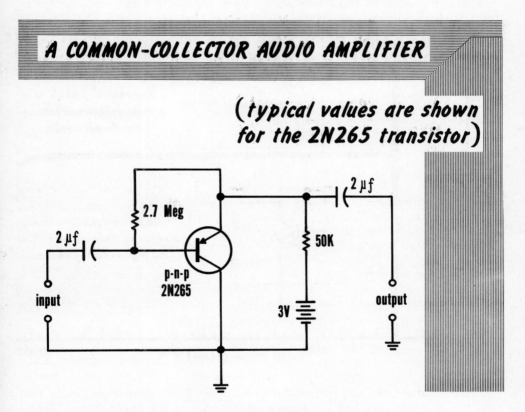

A COMMON-COLLECTOR AUDIO AMPLIFIER

(typical values are shown for the 2N265 transistor)

pected to give power gain in the order of 60 for the common-base circuit and nearly 10,000 in the common-emitter circuit. Although these values are never achieved in practice (since they are based on ideal conditions that are difficult, if not impossible, to realize), power gains in both these circuits are quite high. In general, in a common-collector circuit, the power gain may come to about half the value obtained in the common-base arrangement. One wonders, then, where a common-collector amplifier could be profitably used.

Using the Common-Collector Amplifier

The low voltage and power gains available from a common-collector amplifier indicate that one would seldom find this configuration used under conditions that call for reasonable amplification. Suppose, however, that one wanted to use it as an *impedance transformer*. Of what use would it be in this connection?

A crystal microphone normally has a high impedance and may therefore be coupled directly to the high impedance vacuum tube grid circuit without using a transformer or any other impedance-matcher. This is not good practice, however, in transistor work. Both high-gain transistor configura-

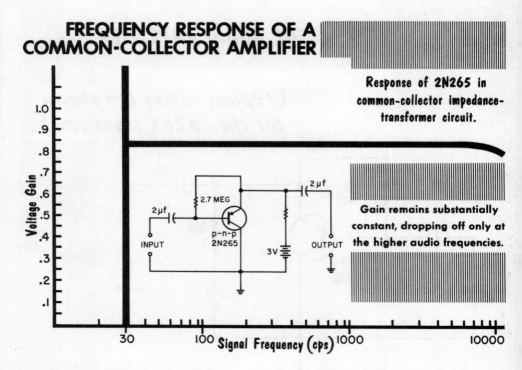

FREQUENCY RESPONSE OF A COMMON-COLLECTOR AMPLIFIER

Response of 2N265 in common-collector impedance-transformer circuit.

Gain remains substantially constant, dropping off only at the higher audio frequencies.

tions (common-base and common-emitter) have *low* input impedances that would represent very bad impedance mismatches to the microphone. Transformers can be used to improve the impedance match, but are generally undesirable because of frequency response considerations. (A transformer of the turns ratio needed for this job is very difficult to manufacture; its frequency response has to be flat over the whole audio range). Instead, a common-collector amplifier may be used between the microphone and the succeeding high-gain common-emitter stage as an impedance transformer with excellent frequency characteristics. Using the circuit values given in the previous diagram, we obtain the frequency response curve shown.

Comparison of Frequency Responses of Configurations

Certainly one of the more important considerations in the selection of an amplifier circuit is that of frequency response. We have just seen that the common-collector configuration is capable of excellent frequency response. Let us compare the response curves of all the circuits on a single graph so that we can arrive at a reasonable conclusion relative to their applications. These curves were obtained with the same transistor in a circuit containing properly matched components. The common-collector curve at once shows itself superior to the others in sheer range of uniform response over the audio spectrum. The poorest response is produced by the common-emitter

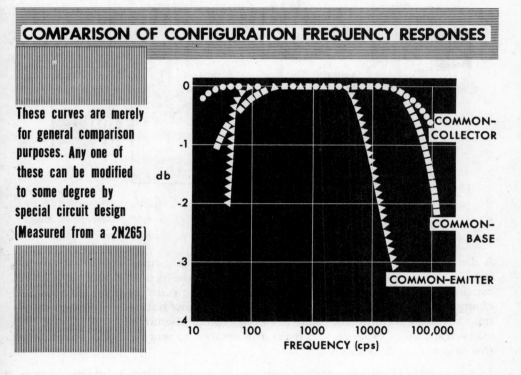

COMPARISON OF CONFIGURATION FREQUENCY RESPONSES

These curves are merely for general comparison purposes. Any one of these can be modified to some degree by special circuit design (Measured from a 2N265)

circuit, while a response somewhere in between these two extremes appears characteristic of the common-base configuration.

From the point of view of popularity with designers and manufacturers, there is no question but that the common-emitter circuit is a favorite. Thus, in all but the highest-fidelity equipment, voltage and power gain are considered of greatest importance. Where gain is not as important as extended frequency response, the common-base circuit is favored. Finally, for impedance transformation, with excellent frequency characteristics, the common-collector circuit is usually chosen.

Sample Circuits — A D-C Amplifier

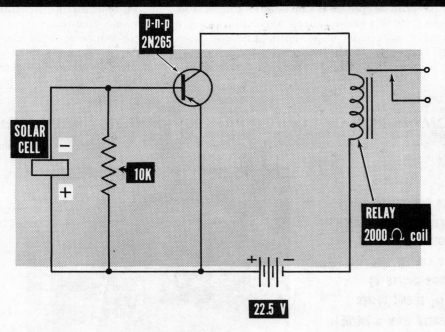

A D-C AMPLIFIER SUITABLE FOR OPERATING A STANDARD RELAY FROM A SMALL SOLAR CELL

p-n-p
2N265

SOLAR CELL

10K

RELAY
2000 Ω coil

22.5 V

A d-c amplifier is called for when we want to build up a small d-c current to a larger value. In general, the function of the d-c amplifier is that of causing a small change of current in its input circuit to produce a larger *change* of current in its output circuit. This immediately suggests that we take advantage of the fact that the *beta* (common-emitter current gain) of many transistors is large. To make the beta go to work for us, we will use the common-emitter circuit.

In our sample circuit we have shown, as the source of input current change, a small solar cell which can produce a tiny current when light shines upon it. Even in the dark, a tiny residual photocurrent still flows, giving the base a small, but adequate, forward bias. Now when light strikes the photocell, the forward base current increases, causing the collector current to rise by a much greater amount. Although the full beta of the 2N265 (beta = 110) cannot be realized in a simple circuit of this variety, a *practical* gain of more than 50 is possible. With the relay correctly chosen, a current rise in the base circuit of 0.1 ma can give rise to a change of 5 ma in the collector circuit. This is more than enough to pull-in a good 2000-ohm relay.

Sample Circuits — An Audio Amplifier

Let us look now at a well-designed, typical audio amplifier. This circuit uses a 2N190 p-n-p transistor with voltage-divider base bias in a common-emitter circuit. It also contains an emitter stabilizing resistor. With the values of resistances given, a collector current of 1 ma flows with zero signal input. Thus, there is a voltage drop of 6 volts across R_1 and R_2 in series in the collector circuit, leaving 6 volts of actual applied potential between the collector and emitter. The input resistance, as seen by the signal generating device, is approximately 1100 ohms; this is of the proper order

A WELL-DESIGNED AUDIO AMPLIFIER
WITH A VOLTAGE GAIN OF 170

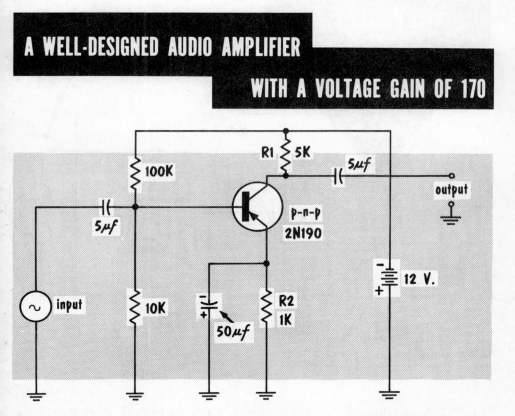

for the common-emitter circuit, as we have seen. The voltage gain of this circuit is very close to 170. Thus, for example, a microphone having peak a-c voltage of 0.01 volt would cause an output voltage swing of about 1.7 volts across R_1, the load resistance.

The use of large coupling and by pass capacitors insures good low-frequency response — down to 20 cycles per second or better. The high-frequency response begins to show a cutoff at about 10,000 cycles.

Sample Circuits — A Simple Radio Receiver

With the advent of transistors has come a degree of miniaturization impossible before, even with miniature vacuum tubes. In this circuit, the transistor is used as an audio amplifier coupled to the output of a diode detector. A radio like this can be fitted into a plastic or metal case no larger than a package of cigarettes, with a hearing aid type of earphone completing what might be called a truly vest-pocket size receiver.

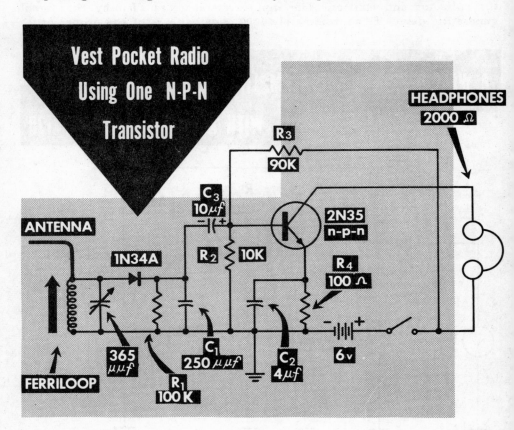

The IN34A rectifies the incoming signal from the high-Q ferrite antenna loop. The demodulation is completed by the diode load resistor R1 and r-f bypass capacitor C_1. The modulation voltage is then passed on to the base of the transistor via the coupling capacitor C_3. R_2 and R_3 comprise the now-familiar bias voltage divider for the common-emitter connection of the transistor. Resistor R_4 is the emitter stabilizing resistor and is by passed by C_2 to prevent degeneration of the signal by R_4. The current drain is so tiny — under 20 microamperes — that the very smallest of penlite cells may be used as the power source.

Sample Circuits — A Transistor Switch

One seldom thinks of a switch as being anything more than a pair of contacts that may be brought together and then separated. In many applications, particularly in computers and related equipment, a switch may be used many millions of times in the course of relatively short intervals. Such switches must carry extremely small currents and must be operated at very low voltages to prevent arcing, oxidation, and possible electrolytic action at the contact surfaces. Any of these effects change the contact resistance and alter the circuit characteristics.

Transistors are now being widely used to serve as transfer devices between the switch and the load as shown in the diagram. It is desired that 25 volts be applied across the load when the switch is closed. Since the load is 125 ohms, then the current that will flow is 200 ma and the power delivered will be 5 watts. The switch, however, is to be operated at only 1.5 volts with as small a current as possible between its contacts. With the switch open, I_{co} is virtually zero so that the load receives no energy. Upon closure, the base is forward-biased, the emitter-to-collector resistance of the transistor drops to 1 ohm, and full power is applied to the load.

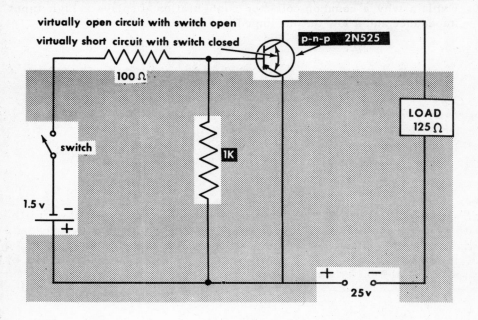

A TRANSISTOR SWITCH

The 2N525 transistor is an alloy junction type whose resistance drops to only one ohm when the base is forward biased, permitting collector conduction.

virtually open circuit with switch open

virtually short circuit with switch closed

p-n-p   2N525

100 Ω

LOAD
125 Ω

switch

1K

1.5 v

25v

QUESTIONS AND PROBLEMS

1. Name the considerations that must be given to the design or selection of an amplifier for a specific use. Explain the importance of each of these considerations.

2. Prove that a common-base amplifier produces a phase inversion of input to output.

3. In a similar manner, show that a common emitter amplifier does not produce such a phase shift.

4. Explain why a high-impedance source (such as a crystal phono cartridge) must be coupled to a common-base amplifier through a transformer. Now explain why such a transformer would not be necessary when the configuration is changed to common-collector.

5. What are the chief advantages of the common-emitter circuit over the other two configurations?

6. Define "beta." For what circuit configuration is beta used?

7. List the input and output impedances for typical common-base, common-emitter, and common-collector configurations. Why is a knowledge of impedance in this connection important?

8. If you wanted to couple the output of a crystal microphone to a low-impedance line, what kind of transistor circuit would you use? Why?

9. What are the factors that cause transistor bias to vary? Why is constant bias important in amplifier operation?

10. Explain why a common-collector configuration displays a high input impedance and a low output impedance.

Coupling Methods Reviewed

AMPLIFIER GROUPS

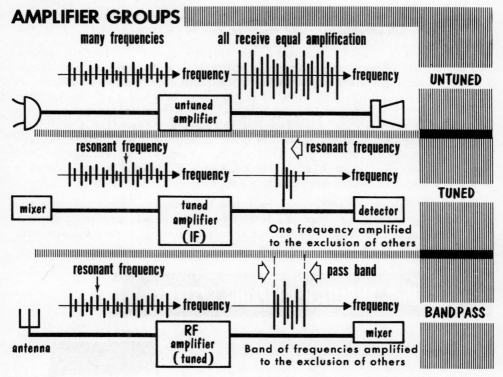

We take advantage of the gain capability of a single vacuum tube amplifier in many simple applications, but when very large amplifications are required we generally couple two or more amplifiers together *in cascade*.

Considering only a-c signal amplifiers (i.e., omitting d-c amplifiers), high gain can be obtained from transistors by using coupling methods very similar to those used for vacuum tubes. We can catalog our thinking a bit better in this direction by recognizing first that we can discuss coupling methods separately as applied to two different groups of systems:

(1) *Untuned amplifiers* such as audio systems in which tuning, as it is usually conceived, does not exist. Such amplifiers are presumed to be flat in frequency response since they should not have greater gain for one frequency than another.

(2) *Tuned amplifiers* as encountered in r-f and i-f systems. Here a single frequency or a narrow band of frequencies is to be given preference over all others. Wideband or bandpass amplifiers that handle many adjacent frequencies over a more or less greater range are also included as a special case of tuned amplifiers.

Coupling Methods Reviewed (contd.)

Untuned coupling systems may be further subdivided into three major classes, although combinations of these classes have been used.

One of the most familiar of these is transformer coupling. In this system, the a-c signal current flows through the primary winding of the transformer, inducing a voltage of the same frequency across the terminals of the secondary winding. The induced voltage may be greater or smaller than the primary voltage, depending upon whether the transformer has a stepup or stepdown turns ratio. Also familiar is resistance-capacitance coupling. In this arrangement, signal currents flow through R_1, producing voltage drops at the same frequency as the current. The varying voltage is then applied across R_2 through coupling capacitor C, where it appears as a voltage across the output terminals.

Coupling resistors are sometimes replaced by coupling impedances in the form of coils to make up what is known as an *impedance-coupling* network. The advantages and disadvantages of a coil in lieu of a resistor for coupling purposes will be discussed later.

UNTUNED AMPLIFIER COUPLING SYSTEMS

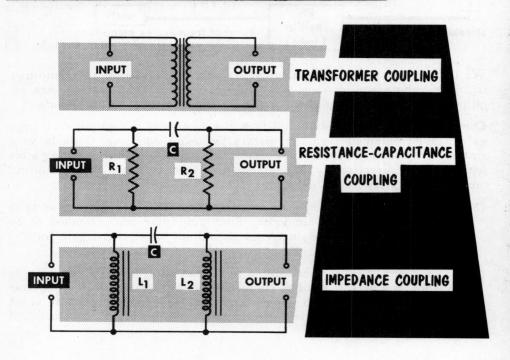

Coupling Methods Reviewed (contd.)

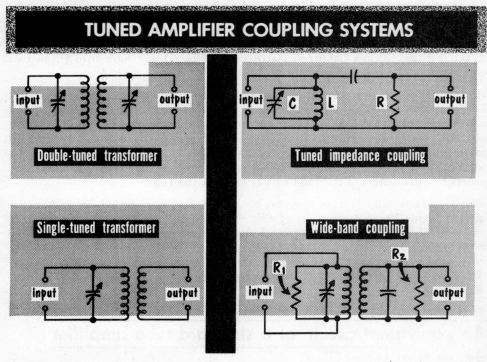

TUNED AMPLIFIER COUPLING SYSTEMS

Double-tuned transformer

Tuned impedance coupling

Single-tuned transformer

Wide-band coupling

Sharply tuned amplifiers, as in i-f stages or in multistage transmitters, are most often coupled by a double-tuned transformer. In the interest of economy, only one coil of the transformer may be tuned while the other is designed to resonate close to the desired frequency without additional tuning.

A single resonant circuit is often used as an impedance coupler very similar to the system used in impedance coupling in untuned amplifiers. In this arrangement the coil L and the capacitor C are adjusted for the desired frequency. Thus, this frequency is the only one that produces a large voltage drop across the terminals of the resonant circuit. Energy is transferred to the output by the coupling capacitor.

The three coupling methods just described are all sharply tuned, i.e., responsive to a narrow band of frequencies close to the resonant point. There is a great deal of demand in modern electronics, however, for bandpass amplifiers, or amplifiers that tune broadly over an extended range of frequencies. Broadening the response may be handled in several ways. One of the most common is the use of loading resistors across one or both of the transformer coils. This reduces gain but widens the amplifier's passband.

Functions of a Coupling System

A coupling system of any type has a very specific job to do: it must transfer the signal voltage or power from the output of one amplifier to the input of another. This it must do with a minimum of loss and without upsetting the voltage balance of either amplifier. Let's look into these two conditions.

In vacuum tubes some amplifiers do not transfer power into the next stage. For instance, a simple voltage amplifier may cause the grid voltage of the following stage to vary about its fixed bias point without supplying any current worth mentioning. In such a coupling system, we must be careful to provide the first tube with a load impedance large enough, (compared to the plate resistance of the tube), to insure that most of the amplified voltage appears across the load rather than in the tube.

The standard tube circuit may be simplified in the form of an equivalent circuit to facilitate discussion of this effect. The tube is really a generator and is so shown in the equivalent circuit; it has a definite internal resistance with normal bias so that this is shown as R_{gen} in the diagram. This resistance is in series with the total load that the tube "sees". If the coupling capacitor C is large enough (low resistance), the tube will look into a load consisting of R1 and R2 in parallel.

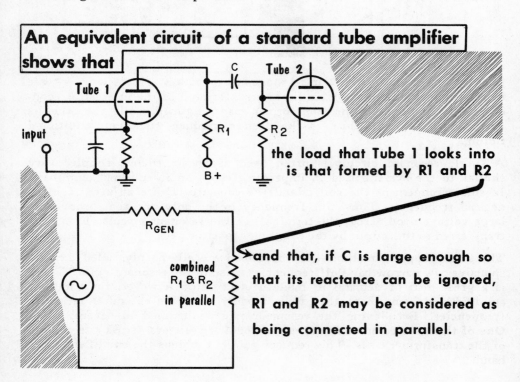

An equivalent circuit of a standard tube amplifier shows that

the load that Tube 1 looks into is that formed by R1 and R2

and that, if C is large enough so that its reactance can be ignored, R1 and R2 may be considered as being connected in parallel.

Functions of a Coupling System (contd.)

THE VALUE OF THE PLATE LOAD RESISTANCE IS IMPORTANT IN COUPLING

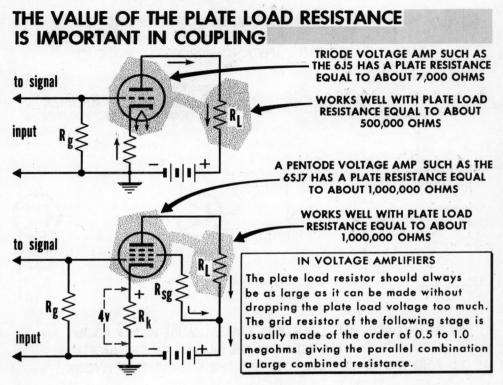

TRIODE VOLTAGE AMP SUCH AS THE 6J5 HAS A PLATE RESISTANCE EQUAL TO ABOUT 7,000 OHMS

WORKS WELL WITH PLATE LOAD RESISTANCE EQUAL TO ABOUT 500,000 OHMS

A PENTODE VOLTAGE AMP SUCH AS THE 6SJ7 HAS A PLATE RESISTANCE EQUAL TO ABOUT 1,000,000 OHMS

WORKS WELL WITH PLATE LOAD RESISTANCE EQUAL TO ABOUT 1,000,000 OHMS

IN VOLTAGE AMPLIFIERS

The plate load resistor should always be as large as it can be made without dropping the plate load voltage too much. The grid resistor of the following stage is usually made of the order of 0.5 to 1.0 megohms giving the parallel combination a large combined resistance.

The tube and its load effectively form a series circuit. Then, as the input signal to tube 1 causes the plate current to vary, there will be two voltage drop variations, one in the internal tube resistance (R_{gen}) and one in the combined load. Since the voltage variations across the load are the ones that are transferred to tube 2, we want these to be as large as possible. Any voltage changes that occur in the tube itself are wasted as far as transfer is concerned.

In a series circuit, the current is the same throughout, hence any voltage drop that appears must be proportional only to the resistance causing it. This would seem to dictate that the combination of R1 and R2 ought to be made as large as possible because most of the signal voltage would then appear across it. There is one drawback to this reasoning, however. The plate voltage of tube 1 is delivered to this tube through R1. If R1 is too large, the plate voltage of the tube will be too small and the tube will then operate on the wrong part of its characteristic curve. For this reason, the choice of R1 is usually a compromise: large enough to provide a reasonably large voltage drop and small enough to allow the plate voltage to remain at the correct level.

Functions of a Coupling System (contd.)

BOTH POWER AMPLIFIERS IN TUBE CIRCUITS AND ALL TYPES OF TRANSISTOR AMPLIFIERS MUST HAVE THEIR OUTPUT AND INPUT IMPEDANCES PROPERLY MATCHED

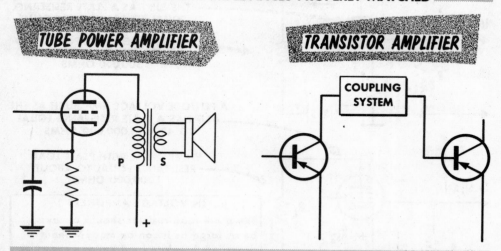

TUBE POWER AMPLIFIER

TRANSISTOR AMPLIFIER

COUPLING SYSTEM

IN BOTH CASES, THE COUPLING SYSTEM MUST BRING ABOUT AN IMPEDANCE MATCH BETWEEN THE OUTPUT OF THE AMPLIFIER AND THE LOAD THAT FOLLOWS THE COUPLING NETWORK.

The rule developed on the previous page is equally applicable to transformer and impedance coupling: the load impedance should be as high as voltage conditions permit.

Suppose, now, that we consider a *power amplifier*. Such an amplifier must supply current as well as voltage to the device that follows it. This may be a speaker, another power amplifier, a modulator transformer or a relay. In any case, we now must consider *impedance matching*. As we know, maximum power transfer from driver to driven stage takes place when the output impedance of the driver matches the input impedance of the driven stage. Since transistors are essentially *current-operated* amplifiers, we shall find that impedance matching in transistor circuits is much more important than in voltage amplifier tube circuits.

Although impedance matching is most easily accomplished through the medium of a matching transformer, it is often more convenient and economical to use resistance-capacitance coupling between transistor stages. Thus, we shall have to give careful attention to impedance matching problems as they arise in our coupling networks.

Functions of a Coupling System (contd.)

There is one additional function of a coupling system that should be reviewed at this point. This has to do with the d-c voltage distribution to both amplifiers.

Going back to tubes for a moment, you will remember that we cannot connect the plate of the first tube directly to the grid of the second, since this would apply the positive plate potential to the grid, too. Since an amplifier grid is generally at a negative d-c potential with respect to the cathode, this would set up hopelessly incorrect bias conditions. Either a transformer or a capacitor will block d-c without blocking the a-c component of the signal at the same time. Thus, any coupling system must be arranged with this matter of voltage distribution in mind. *The coupling system must not alter bias conditions that each individual stage demands.* In other words, the coupling system must be a transferal agent for a-c, but must block the d-c if this blocking action is called for. As we shall see later, certain transistor circuits may be coupled directly, making a capacitor or transformer unnecessary. This cannot be done with tubes without elaborate voltage adjusting networks.

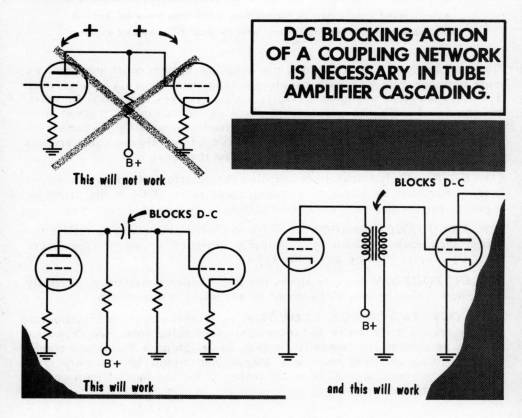

D-C BLOCKING ACTION OF A COUPLING NETWORK IS NECESSARY IN TUBE AMPLIFIER CASCADING.

This will not work

BLOCKS D-C

This will work

BLOCKS D-C

and this will work

Objectives in Selection of Coupling System

	GAIN	FREQUENCY RESPONSE	LACK OF DISTORTION	ECONOMY
BEST				
INFERIOR				
MORE INFERIOR				

COMPARISON OF COUPLING CRITERIA

The chart does not imply that you cannot get good frequency response with well-designed transformers or impedances; it indicates the good features obtainable from the coupling systems with the least difficulty and expense.

Now that we see what functions the coupling system must perform, we must now ask *how well* each coupling arrangement does its various jobs. This question leads to another: exactly what objectives do we look for when we are faced with the problem of choosing one coupling method over another? There are actually four major criteria for passing judgement on the performance of a multistage amplifier. To a large degree, these criteria are dependent upon coupling. Let us see what they are:

(1) *GAIN* — Only transformer coupling can contribute to the voltage gain of an amplifier by virtue of a stepup turns ratio. This is the principal reason for the popularity of this method.

(2) *FREQUENCY RESPONSE* — By and large, it is easier to realize the best frequency response by a properly designed resistance-capacitance (R-C) coupler than by any other type.

(3) *DISTORTION* — Here again, the R-C coupling method is generally the best, unless one cares to invest in expensive transformers.

(4) *COST AND SPACE ECONOMY* — Transformers and audio impedances are more costly and more space-consuming than R-C networks. Thus, very often it is actually cheaper, in designing a multistage amplifier, to plan on using more R-C stages than would be necessary with transformer coupling. This would make up for the loss of gain contributed by transformers.

A Transformer-Coupled Transistor Amplifier

When two 2N265 audio transistors are transformer-coupled in cascade, a total power gain of close to 90 db is obtained provided that the transformers are carefully selected for impedance matching characteristics. The input impedance of the 2N265 is approximately 4,000 ohms in the common-emitter circuit; its output impedance is close to 16,000 ohms. Thus, T1 should have a stepdown turns ratio that provides this impedance ratio. With well-built transformers, this does not affect the *power* transfer from primary to secondary. T2 must have a primary impedance in the vicinity of 16,000 ohms but its secondary must be selected to match the load that is to be used. For instance, if the output is to work into a speaker, the secon-

TRANSFORMER-COUPLED TRANSISTOR AMPLIFIER WITH A GAIN OF CLOSE TO 90 DB

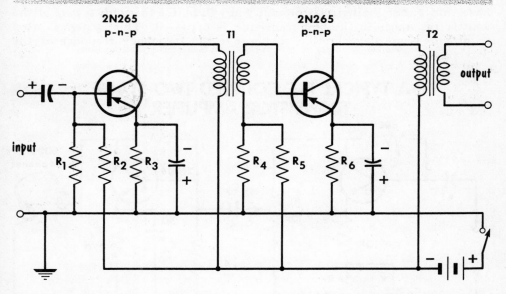

dary might have an impedance of 3 to 4 ohms; if the load is to be a pair of magnetic headphones, the transformer secondary might have an impedance of 1000 to 2000 ohms.

Resistors R1 and R2 form the voltage divider network needed to provide the correct base bias. As previously emphasized, the transformer isolates the transistors with regard to d-c bias voltages, hence each one must be individually biased. R3 is the stabilizing emitter resistor, bypassed to prevent degeneration. This circuit provides high gain; the frequency response, distortion and cost will depend upon the quality of the transformers.

An R-C Coupled Transistor Amplifier

For comparison with the transformer-coupled amplifier just discussed, study this R-C coupled amplifier. Originally designed as an audio preamplifier for use with a crystal microphone or phono pickup; it may be used for any general audio amplification purposes where a power gain of 50 db is satisfactory. The power output is about 30 milliwatts, adequate for driving a 2-watt class-A power amplifier or an 8-watt class-B amplifier. These will be discussed under the subject of power amplifiers.

Note the R-C coupling system consisting of R4, C3, and the combination of R5 and R6. The signal voltage drop for transfer to the second stage appears across R4. Since this resistor is connected to the + terminal of the battery, C3 is required to prevent this potential from reaching the base of the next transistor; in other words, as in vacuum tube R-C circuits, it is a *blocking* capacitor as well as a coupler. The signal current now flows into the base circuit of the second transistor to be further amplified. R6 and R5 make up the now-familiar base-bias voltage divider. Both transistors have emitter stabilizing resistors but you will notice that *only a part* of the emitter resistance in the first transistor, is bypassed. This system is used to provide some degeneration to improve the frequency response of the first stage.

A TYPICAL R-C COUPLED TWO-STAGE TRANSISTOR AMPLIFIER

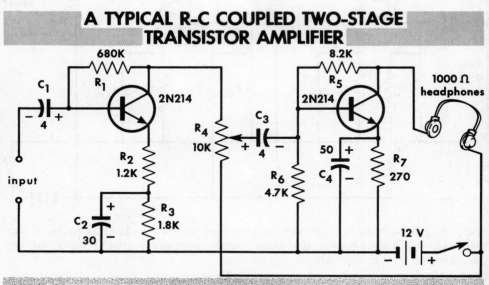

NOTE THAT THE GAIN OF THIS SYSTEM IS ABOUT 50 db. This is 40 db less than the transformer coupled system previously discussed. The reduced gain is partially due to the fact that 2N265 transistors have a higher beta, and partially due to the fact that TRANSFORMER COUPLING ALWAYS GIVES MORE GAIN THAN RESISTANCE COUPLING FOR THE SAME TRANSISTORS.

Combination Coupling

Let us now examine a commercial cascaded transistor amplifier used as a telephone pickup device. A special flat coil is placed under the telephone. The signal currents flowing in the telephone inductances induce currents of similar frequency in the coil. These induced currents are then amplified sufficiently so that the conversation may be heard on a pair of headphones, or may be used to make a tape recording.

Three transistor stages are utilized. The first is coupled to the second through a resistance-capacitance network while the second is coupled to the third via a matching transformer. The biasing is of the simplest variety — merely a resistor that joins the base to the minus terminal of the bat-

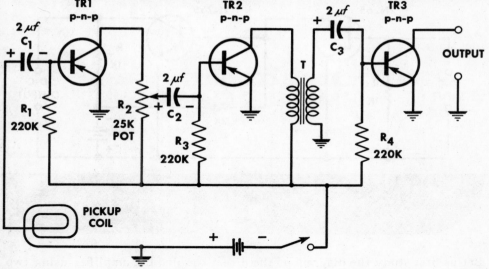

AN AMPLIFIER WITH COMBINED TRANSFORMER AND R-C COUPLING

tery. No emitter stabilization is used since high fidelity is not required, nor need the frequency response be particularly flat. The frequencies reproduced by the telephone cover a narrow range. Thus, a small shift along the transistor characteristic curve due to temperature changes would have little effect upon the performance of this unit.

Transformer T is selected to match the output impedance of TR-2 to the input impedance of TR-3. The use of coupling capacitor C_3 is a little unusual, but in this connection (called *shunt feed*) the base of TR-3 is isolated from ground, *permitting R4 to develop the proper bias*.

Direct Coupling — Common Emitter to Common Emitter

In our earlier discussion on coupling systems (p. 105) it was pointed out that direct coupling in vacuum tube circuits was seldom used because of the biasing problems it causes. It was also mentioned that it *is* possible to couple transistor amplifiers directly, provided that precautions are taken to maintain correct d-c voltage conditions. This is considerably easier to do with transistors than it is with vacuum tubes.

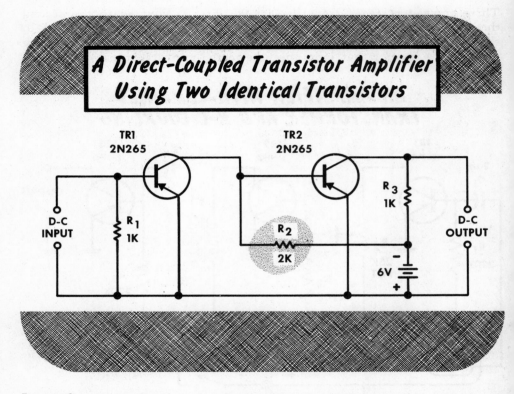

A Direct-Coupled Transistor Amplifier Using Two Identical Transistors

Let us first study the diagram of the direct-coupled d-c amplifier using two 2N265 p-n-p transistors. How are the voltage conditions adjusted to satisfy the demands of the transistors relative to proper bias? Reviewing the bias requirements of p-n-p units, let us write it this way:

Collector	— —
Base	—
Emitter	+

This merely means that the collector is more negative than the base and that the base is more negative than the emitter. These conditions *must be fully met* in any standard amplifier we build. We can start our analysis by examining the events that take place in and around R2.

Direct Coupling — Common Emitter to Common Emitter (cont.)

SHOWING HOW VOLTAGES MAY BE ADJUSTED FOR PROPER BIAS IN A DIRECT-COUPLED AMPLIFIER

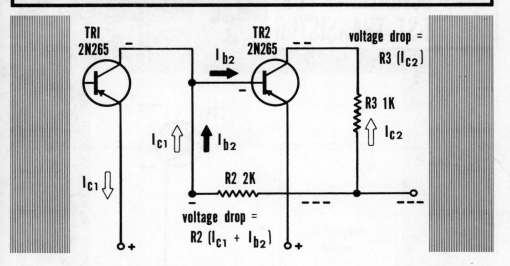

Since R2 is common to the collector of TR1 and the base of TR2, the voltage drop across it will be determined by the product of R2 times the *sum* of I_{c1} and I_{b2} as shown in the diagram. The potential on the right side of R2 is the most negative point in the circuit since this terminal is connected to the negative pole of the battery. Let us label this point with three signs (− − −) as shown. Since two currents flow through this 2000-ohm resistor, there will be a sufficiently large voltage drop across it to justify our label of a single minus sign (−) on the left side of R2. Thus, the base of TR2 is negative with respect to the emitter. We have established thus far the condition:

Base	−
Emitter	+

Only one current flows through R3, however. Assuming that it is of the same order as I_{c1} (since both transistors are the same) and noting that it is half the resistance of R2, the voltage drop across it must be smaller than the drop across R2. Hence, the potential of the collector of TR2 must be more negative than the base of the same transistor. This gives us the condition:

Collector	− −
Base	−

This meets all the biasing requirements discussed before.

Direct Coupling — Common Emitter to Common Emitter (cont.)

BIAS CONDITIONS FOR THE FIRST TRANSISTOR

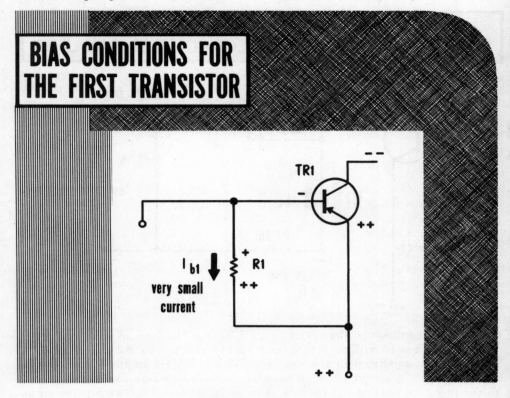

As you looked at the diagram on the preceding page, it might have occurred to you that the collector of TR1 is no longer at a negative potential as high as that of the collector of TR2. Will this disturb the bias conditions for TR1? Not at all.

Think of it this way: the collector of TR1 is definitely negative with respect to the common ground of the system, although it may not be quite as negative as the collector of TR2. The base current of any transistor is very small, as we have seen many times before. Therefore, the voltage drop across R1 will also be quite small, leaving the base just *slightly more negative* than the emitter. As we have shown it in the diagram, the base is slightly *less positive* than the emitter, which is the same thing. Thus, for TR1 the collector is definitely negative (and we symbolize this as − − for this transistor considered alone), the base is less negative than the collector and the emitter is more positive than both. In short:

Collector	− −
Base	−
Emitter	+

Complementary Nature of Transistors

There is only one *variety* of vacuum tube. That is, any tube, regardless of the number and kinds of electrodes it contains will always have an electron flow from cathode to plate. The plate is always positive with respect to the cathode. Transistors, however, come in two distinct varieties: the p-n-p and n-p-n. Although their respective performances are very nearly the same in nature, they are designed to operate with oppositely directed bias potentials. These potentials may be summarized thus:

	p-n-p	*n-p-n*
Collector	− −	+ +
Base	−	+
Emitter	+	−

This oppositeness of bias polarities is referred to as a *complementary condition*. It opens up an entirely new approach to coupling. As we shall show, it makes it possible to direct-couple two transistors without worrying at all about adjusting bias voltages. This means a saving of parts and space, as well as higher gain and fewer circuit complications.

Let us first couple two transistors in this complementary fashion and see what happens to the bias voltages.

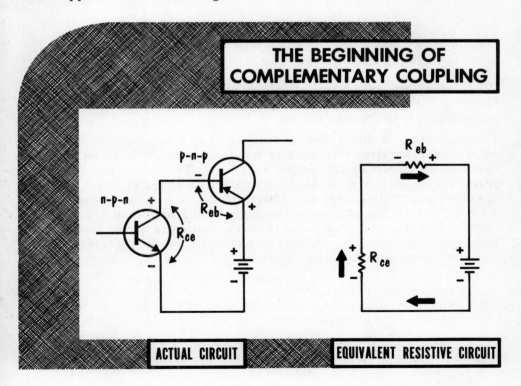

THE BEGINNING OF COMPLEMENTARY COUPLING

ACTUAL CIRCUIT

EQUIVALENT RESISTIVE CIRCUIT

Analyzing the Coupling

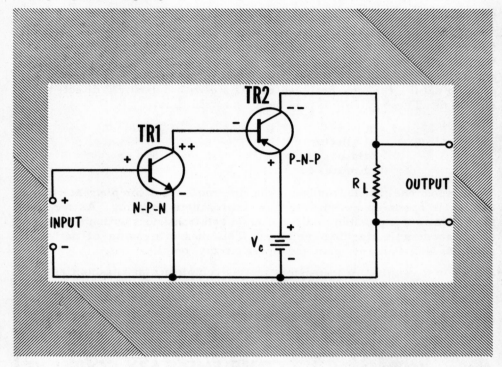

We have shown two common-emitter configurations in complementary coupling but have omitted the input and output circuits to clarify the relationships in the coupled section.

Complementary coupling may be thought of as a condition in which the emitter-collector circuit of the first transistor (left) is in series with the base-emitter circuit of the second transistor. The equivalent circuit replaces the transistors with resistances. Since these are in series, the voltage drops around the circuit would have the polarities shown in the diagram. Now we transfer the plus and minus signs to the equivalent points on the transistors and examine the biases to check their correctness. Listing these and comparing them with the summary on the previous page, we find that they are just as they should be. Thus:

	p-n-p	n-p-n
Collector		+
Base	−	
Emitter	+	−

Unless we add some form of output load and input voltage, we cannot tell what the remaining electrode biases will be. Let us do this.

Adding Input and Output

We have fashioned a d-c amplifier by adding some form of d-c input voltage (say from a photocell or a thermocouple) and an output load resistor. The diagram shows the remaining electrode voltages that appear as a result of the input and output circuits.

TR2 is forward-biased by the common battery, V_{cc}. Thus, the resistance of the emitter-base circuit of TR2 is quite small and the voltage drop across it, due to the series-circuit current, is very small. This means that the positive voltage applied through TR2 to the collector of TR1 is almost the same as that of the battery. For this reason, we have taken the liberty of marking the collector of TR1 with the + + sign; this merely indicates a comparatively large positive potential.

Looking at the circuit from the other direction, we know that the collector of TR1 is *reverse-biased*, giving this path a high resistance. Electrons from the battery (minus terminal) must, therefore, travel through this high resistance before reaching the base of TR2. The large voltage drop in TR1

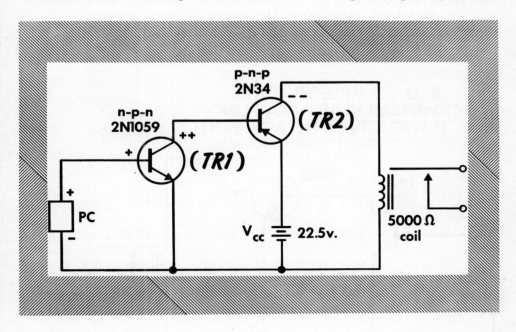

leaves the base of TR2 still negative, as it should be, but *only slightly* negative. On the other hand, the negative terminal of V_{cc} goes directly through R_L to the collector and since R_L may be relatively small, we can mark the collector − −. Finally, the base of TR1 is marked + because the input voltage source would necessarily produce a small voltage that would not drive the base more positive than the collector.

A D-C Amplifier Using Complementary Coupling

A direct outgrowth of the circuit just discussed is that of a photocell amplifier circuit using a 2N1059 n-p-n transistor and a 2N34 p-n-p type in a complementary-coupled circuit. Note how few parts are used. The photocell may be any one of the self-generating types, either a selenium barrier-layer unit or one of the newer silicon solar cells. With no light falling on the photocell, the base current of TR1 is very small, making the collector current of this transistor virtually equal to I_{co}, also very small. Hence, the base current of TR2 is also very small since the collector current of TR1 *is* the base current of TR2. Result: TR2 collector current is too small to operate the relay. With light falling on the photocell we get this sequence: increase in TR1 base current — increase in TR1 collector current — increase in TR2 base current — sufficient increase in TR2 collector current to pull in the relay.

You will be interested in the gain possible in such a circuit. An input power of approximately .05 microwatts yields an output power of better than 10,000 microwatts! This might have more meaning when considered in terms of current gain. Since the current gain of a complementary-coupled circuit using these transistors is approximately 600, a 5-ma relay will be pulled in when the photocell produces only 8 microamperes!

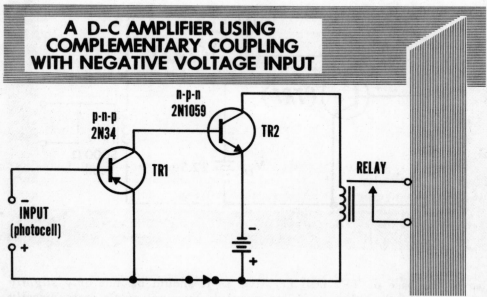

A D-C AMPLIFIER USING COMPLEMENTARY COUPLING WITH NEGATIVE VOLTAGE INPUT

Compare this circuit with the previous one given for a d-c amplifier. The circuits are identical except that the p-n-p and n-p-n transistors are reversed in position. This calls for a battery reversal as well. In this circuit, an increasing NEGATIVE potential input causes the relay to pull in.

An Audio Amplifier Using Complementary Coupling

This circuit will be recognized as being very similar to the d-c amplifier discussed on the previous page. A fixed base-bias is produced by R1. Since these transistors may differ slightly from one another in characteristics, it

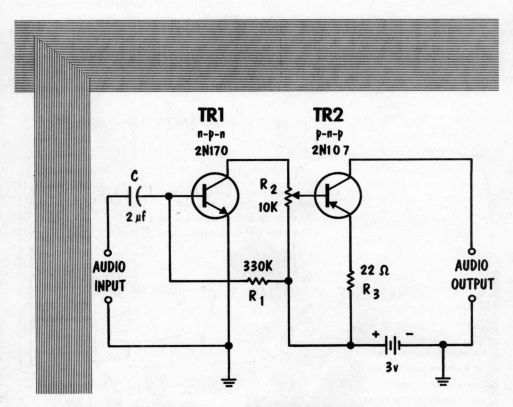

is always advisable to try other values for R1 (without deviating by more than 30% from that given) to produce optimum amplification and fidelity. R2 is a volume control of straightforward type. As the wiper is moved downward, an increasing percentage of the resistance of R2 comes into the series circuit of the complementary-coupled pair. This reduces the audio base current to TR2 and, consequently, the audio currents in the collector circuit.

Note the small value of R3 (22 ohms). A resistor of this size cannot offer much in the way of stabilization but it does act as a safety device, preventing the collector current of TR2 from "running away" if there should be a large increase in temperature. Bypassing is unnecessary; a 22-ohm resistor does not cause sufficient degeneration to warrant the cost of a bypass capacitor.

Power Transistors — What Makes them Different

As we increase the power demands made on a transistor, we must begin to take into account the additional heat being generated in the transistor itself. We have seen that transistors are temperature sensitive. Reliable performance, therefore, depends upon our ability to keep the transistor cool enough, and what is even more important, to maintain it at a constant temperature.

In most applications, a transistor supplies the extra power required by the load in the form of increased collector current. But increased current means more heat generated in the internal resistance of the transistor.

TO SHOWER

An object that is being heated continues to rise in temperature just as long as it cannot get rid of the heat being poured into it.

COOL AIR

An object that can get rid of its heat (conduction, convection, radiation) can remain at the same temperature although heat is being added

WARM AIR

Actually, if the power supplied by a transistor (or any electrical device) is doubled, this means that the amount of heat produced per second must also double, since heat is equal to power *per unit time.* The clue as to how we can prevent a transistor from overheating lies in the time factor. In other words, if we can carry the heat away *fast enough* the temperature of the transistor need not rise without limit even though the power being produced is increased many times over. Thus, the primary prerequisite of a power transistor as compared with a voltage amplifier type such as we have discussed thus far, is that it be able to get rid of heat *quickly and effectively.*

Power Transistors — Physical Structure

Several ways have been used, either singly or in combination, to carry heat away from transistors. Some manufacturers have designed their transistor cases with a ribbed structure to present a large radiating surface area to the atmosphere; this is the same principle as fins on the cylinder of a motorcycle engine. In extreme cases, fans may be installed to increase the rate of heat conduction. Water cooling is another possibility.

WAYS OF KEEPING THE TEMPERATURE OF POWER TRANSISTORS DOWN

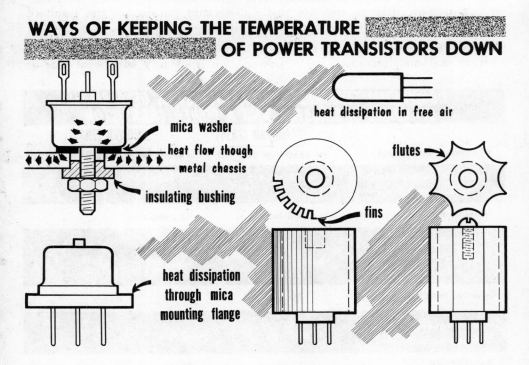

heat dissipation in free air

mica washer

heat flow though metal chassis

insulating bushing

flutes

fins

heat dissipation through mica mounting flange

At the present time, however, the trend in power transistor design seems to have taken the direction of cooling by heat sink. A heat sink may be defined as a relatively large mass of a material having good thermal conductivity. Most common metals, particularly copper and aluminum, are good heat conductors. Thus, the transistor is designed in such a way that the heat producing element — the collector, as a rule — is placed in close contact with the transistor case. The case, in turn, is bolted or riveted to a comparatively large mass of metal, such as a chassis or subchassis which then serves as the heat sink.

In such power transistors, the collector is necessarily at the same potential as the case of the unit and the heat sink as well. This requires electrically insulating the heat sink from the common ground of the system in all circuits involving common-base and common-emitter configurations.

Power Transistors — Typical Characteristics

As we look over the chart of electrical characteristics for a typical power transistor, several differences between these ratings and those given for small transistors become obvious. Both sets of characteristics have been listed side by side for easy comparison.

First, the power transistor can handle better than 65 times the power in its collector circuit, than the standard 2N34 transistor. Note that the maximum collector currents, too, are vastly different: 100 ma for the standard transistor and 3 *amperes* for the power unit.

Second, the current gain of the power transistor is appreciably less than that of the small transistor. The power gain is usually somewhat less as

COMPARING ELECTRICAL CHARACTERISTICS OF A SMALL STANDARD TRANSISTOR AND A POWER TRANSISTOR	STANDARD (2N34)	POWER (2N350)
Maximum dissipation	150 mw (25°C)	10 watts (with heat sink) (80°C)
Maximum collector-to-base voltage at 25°C	-40 volts	-40 volts
Maximum collector current	100 ma	3 amperes
Current gain (beta) minimum	25	20
maximum	125	60
Collector cutoff (I_{cbo})	50 μa	3 ma
Cutoff frequency (common emitter)	10,000 cps	5000 cps
Power Gain (common emitter)	37 db	31 db

well. This is very similar to the characteristics of vacuum tubes; a power amplifier such as a 6V6 does not have much voltage gain (amplification). It is used as an output tube because it is capable of handling large amounts of power. The same is true of the power transistor.

Third, the cutoff current (I_{co}) of the power transistor is much greater than that of its smaller counterpart. Thus, it "idles" with considerable current flowing. This is important from the point of view of battery life.

Finally, the upper frequency limit of the power transistor is quite low, indicating that its use is confined to audio amplification.

Power Transistors — Single-Ended Amplifier

A practical power amplifier having the following ratings is illustrated in the accompanying drawing:

Power output	2 watts
Efficiency	35%
Power gain	25 db
Frequency response	10,000 cycles
Driving power	30 mw

To obtain this power output, the transistor (which is of the case-grounded collector type discussed previously) must be secured to a heat sink at least

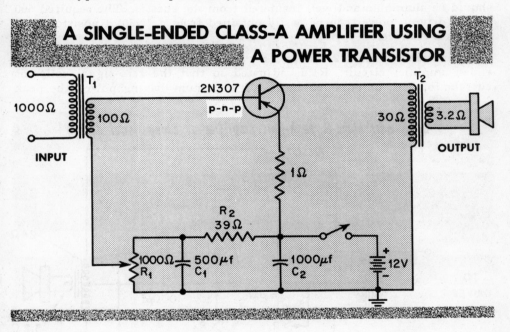

A SINGLE-ENDED CLASS-A AMPLIFIER USING A POWER TRANSISTOR

5×5 inches in area. Aluminum that is 1/16 inch thick is recommended. The heat sink *must be insulated* from common ground.

The input and output impedances of the 2N307 are both quite low and require matching transformers, as indicated. Base bias is provided by the voltage divider consisting of R1 and R2 connected right across the battery. Before placing in operation, R1 should be adjusted for a collector current of 350 ma. This sets the operating point of the transistor right in the center of its linear portion to provide good class-A performance. Because the emitter current is normally high, the emitter stabilizing resistor need be only 1 ohm. Base bias is prevented from fluctuating by C1 and C2, a pair of large capacitors.

Power Transistors — Push-Pull Amplifier

For additional audio power, two power transistors can be connected in a conventional push-pull circuit. If operated class-B as in this case, up to 8 watts of audio power are easily obtained.

Power output	8 watts
Efficiency	60%
Power gain	25 db
Frequency response	8,500 cycles
Driving power	500 mw

The heat sink for the class-B 2N307's should measure $3 \times 5 \times 1/16$ inch, should be aluminum and well insulated from the chassis. The required 500 mw of driving power can be easily obtained from a 2N307 transistor used as a class-A driver.

The functions of R1, R2, R3, and R4 are quite the same as in the single-ended amplifier circuit. R1 is adjusted so that the zero-signal collector current for both units is 100 ma. With maximum driving power, the peak

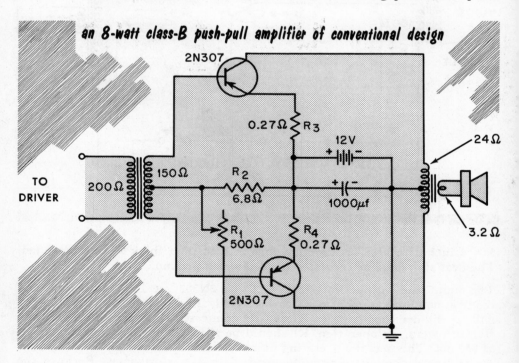

an 8-watt class-B push-pull amplifier of conventional design

collector current is about 900 ma. This magnitude of collector current demands that both transformers be capable of handling rather large currents without overheating.

Power Transistors — Complementary-Symmetry

We have seen how the complementary nature of n-p-n and p-n-p transistors may be utilized in a direct coupling arrangement for cascaded transistors. Now we have enough information to use a pair of complementary transistors in a push-pull circuit that does away with all coupling transformers.

COMPLEMENTARY-SYMMETRY PUSH-PULL CIRCUIT

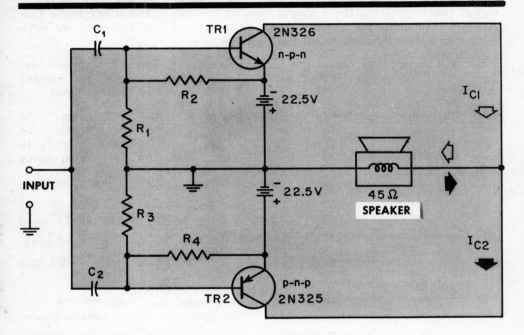

The circuit is satisfactory only if the two transistors have identical electrical characteristics, one being n-p-n and the other p-n-p. Due to the symmetrical character of the transistors, a circuit such as this is known as *complementary-symmetry.*

Let us first examine the circuit action with zero signal input. Resistors $R1$ and $R2$ provide the required base bias for $TR1$; $R3$ and $R4$ do the same for $TR2$. This is standard voltage-divider bias. With perfect symmetry in both collector circuits — and this includes the batteries as well as the collector characteristics — both collector currents will be equal.

Insofar as the voice coil of the speaker is concerned, I_{c1} and I_{c2} flow in opposite directions. Since the currents are equal, the net current in the voice coil is zero. Thus, the voltage drop across the coil is also zero and no power is dissipated in it.

Power Transistors — Complementary-Symmetry (contd.)

Now let us imagine an applied signal that is positive-going. We have simplified the circuit by leaving out the biasing resistors and coupling capacitors. A positive-going voltage applied to the input terminals will drive both bases positive. For the upper transistor, an n-p-n type, an increase in positive base voltage merely causes the forward bias to increase as well; hence the collector current of the upper transistor increases, too. We have shown this by lengthening the white arrow that represents the collector current of the n-p-n transistor. At the same time, the positive-going signal *subtracts* from the forward bias of the lower transistor since this is a p-n-p unit. Its collector current consequently diminishes as we have indicated by the foreshortened dark arrows.

Now observe the currents in the speaker voice coil. There is a larger current to the left than to the right in our picture. Thus, there is a *net current* to the left determined by the difference between the two current flows.

The situation reverses when the input voltage becomes negative-going as shown in the lower diagram. In this case, the net current through the voice coil is to the right. This completes the cycle. Hence, for each input cycle there is a corresponding alternating current flowing in the speaker voice coil to produce the desired reproduction.

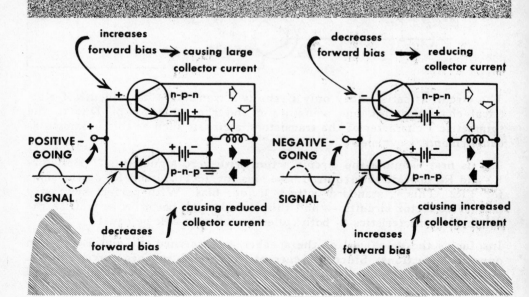

COMPLEMENTARY-SYMMETRY ACTION WITH AN A-C SIGNAL INPUT

High-Frequency Coupling In Practice

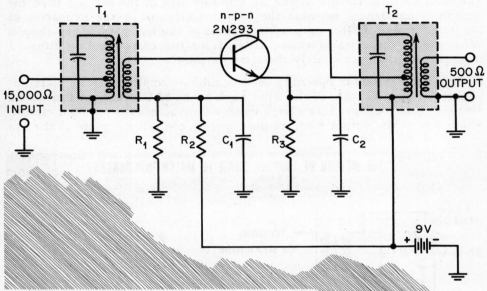

A REPRESENTATIVE I-F AMPLIFIER CIRCUIT USING A HIGH-FREQUENCY TRANSISTOR

Although we have discussed various coupling systems for both high and low frequencies, we have not yet examined a high-frequency circuit in detail. As a good example of modern design practice in this field, let us analyze the circuit of a representative i-f amplifier circuit as it would be found in an up-to-date transistor superheterodyne.

The 2N293 transistor is built for i-f amplification in that its cutoff frequency is in the region of 7 mc. It can be expected to perform well, therefore, at 455 kc (only 0.455 mc), the standard radio intermediate frequency. Coupling from the preceding mixer stage is handled by slug-tuned transformer T1. Similarly, coupling to the detector that follows the i-f stage is accomplished through a similar transformer T2.

Zero-signal collector current is governed by the values of R1, R2, and R3. The first two form the usual base-bias voltage divider, while R3 is a relatively large resistance (of the order of 500 ohms) in series with the emitter. Capacitor C2 bypasses R3 for the conditions that prevail when a signal is received. Without C2, there would be the danger of excessive degeneration resulting in a significant loss of gain. The remaining capacitor, C1 maintains the bias voltage on the base at a constant value so that the transistor will always operate on the linear portion of its characteristic.

High-Frequency Coupling In Practice (contd.)

The most unusual feature of the i-f amplifier lies in the method used for matching impedances between the resonant circuits in the primaries of the transformers and the output impedances of the transistors they couple. Why should these transformers differ from the corresponding tube i-f transformers, used for exactly the same purpose?

A high-gain pentode is generally used in tube i-f amplifiers. The plate impedance of a typical tube of this kind, such as the 6AU7, is about 1 megohm. Since the tube must produce a large voltage drop across the resonant circuit, which acts as the output load, we need merely connect the plate to the *top*

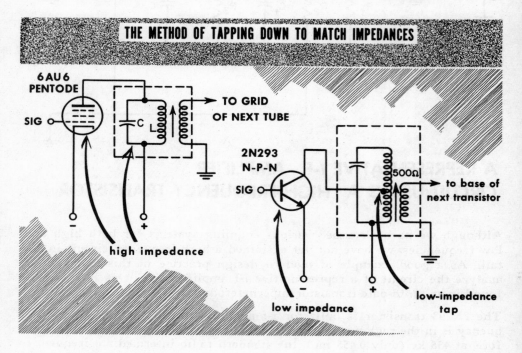

THE METHOD OF TAPPING DOWN TO MATCH IMPEDANCES

of the transformer primary to bring about a good impedance match. It will be recalled that the impedance of a parallel resonant circuit such as that formed by L and C is very high and "goes well" with a plate resistance of 1 megohm.

The output impedance of a 2N293 is given as 18,000 ohms. Since this is substantially less than the impedance of the entire resonant circuit, a much better match can be obtained by tapping down on the primary coil by the correct amount. In addition to this, the input impedance to the next stage — and this might be a diode detector — should be kept low. This accounts for the 500-ohm output winding.

QUESTIONS AND PROBLEMS

1. Discuss the essential differences between the coupling methods used for untuned amplifiers.

2. What are the principal methods used for coupling tuned amplifiers? What are the advantages of each?

3. Why is impedance matching between voltage amplifier transistors more important than between voltage amplifier vacuum tubes?

4. Describe fully the functions of any cascade coupling system.

5. What are the broad objectives in selecting a coupling arrangement?

6. Transformer-coupled transistor amplifiers yield the greatest amount of gain. Why are not all cascaded transistor amplifiers transformer-coupled?

7. Draw a typical R-C coupled amplifier and explain the function of each part.

8. What problems are encountered in attempting to couple two identical transistors? What steps must be taken to overcome this trouble?

9. Why is a p-n-p transistor said to be the "complement" of an n-p-n transistor? How is this characteristic utilized in special coupling systems?

10. Draw the circuit of a complementary-coupled two-stage transistor amplifier for audio purposes, using a p-n-p as the input transistor and an n-p-n as the output transistor. Label each part with the proper approximate magnitude. Finally, explain how it operates.

11. How do power transistors differ in mechanical structure from the smaller "voltage amplifier" types?

12. What is meant by a heat sink? Is a heat sink mandatory whenever power transistors are used? Explain.

13. With the aid of a diagram, explain the operation of a complementary-symmetry power amplifier. Why is an output transformer unnecessary in this circuit? Does the speaker used in complementary-symmetry circuits differ in any way from the speaker found in straightforward push-pull amplifiers? How? Explain.

Sustained Oscillation

An electronic device capable of producing sustained oscillation is always based upon a set of very definite characteristics. Consider a simple vacuum tube Hartley oscillator.

(1) *Frequency-Determining Network.* Oscillators yield a self-maintained a-c voltage at a given *frequency.* This demands the presence of components that set up the conditions needed to establish the oscillation frequency. In the circuit shown, the oscillatory components are L and C; other types of oscillators may use R-C components for frequency control.

(2) *Amplification.* Without amplification, an oscillator cannot sustain its output, except in certain unusual types. Amplification may be provided by a tube or a transistor.

(3) *Positive Feedback.* Some form of positive feedback is required. A portion of the amplified oscillatory voltage is returned to the frequency-determining network to make up for its resistance and radiation losses.

(4) *Automatic Bias.* Most oscillators contain components that bring about the correct bias to sustain the output. Such a bias system must permit easy starting of oscillation and should adjust itself automatically to maintain constant amplitude of oscillatory voltage.

A SIMPLE HARTLEY OSCILLATOR USING A VACUUM TUBE AMPLIFIER

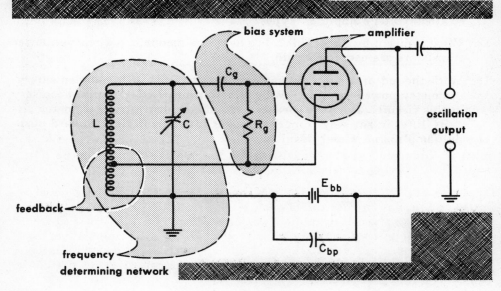

A Transistor Audio Oscillator

Let us see how the oscillator criteria set up on the previous page match this simple transistor audio oscillator circuit.

(1) *Frequency-Determining Network* — Transformer winding L in series with the emitter and the accompanying capacitor (C) "tune" this oscillator since they form an oscillatory circuit of conventional nature.

(2) *Amplification* — The transistor is wired as a common-base amplifier having considerably more impedance (L') in its collector circuit than in its emitter circuit. Thus, it is capable of a definite voltage gain.

the transistor equivalent of a tickler type of self-maintained oscillator.

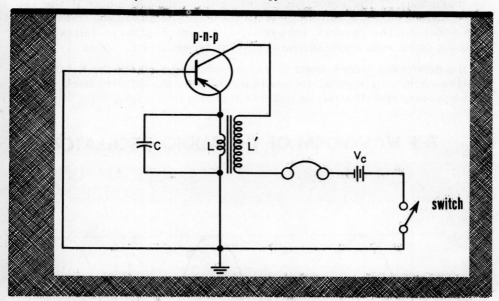

(3) *Positive Feedback* — Coil L' is the tickler winding. Its terminals must be connected so that it will cause positive feedback into the low impedance winding as described on the next page.

(4) *Automatic Bias* — As we know, when we deal with transistors we are interested in bias *current* rather than bias voltage. With proper phasing of the two transformer windings, an induced current into the primary L can be made to flow in a forward direction. Unlike a vacuum tube, the transistor in an oscillator circuit will not necessarily see an unchanging average bias, since the circuit does not contain the equivalent of the bias capacitor in the tube circuit.

Operation of the Audio Oscillator

At the instant that the switch is closed, the base of the transistor is short-circuited to the emitter through L. For this condition, the collector current is merely I_{co}, the collector cutoff current. This is a very small current, but it does grow quickly when the circuit is first completed. Flowing through L' as it does, it induces a voltage in the primary winding L. Assuming that the coils are correctly phased, this new voltage produces a *forward bias* in the base-emitter circuit that causes the collector current to increase from I_{co} to some slightly larger value.

Such a collector current increase must, therefore, raise the forward bias in the base circuit by electromagnetic induction in the transformer, bringing about a further growth of collector current. Swiftly, I_c rises from A toward point B. As it approaches B, the transformer begins to saturate and the magnitude of the induction begins to diminish. At point B the collector current has stopped rising. With the magnetic flux in the transformer core constant, the *induced* base-emitter current begins to fall at once, bringing down with it the collector current as well.

On the downward sweep, there is a small overshoot where the value of I_{co} is driven below its normal quiescent magnitude. As it rises back toward A, the process repeats itself as it did at the start.

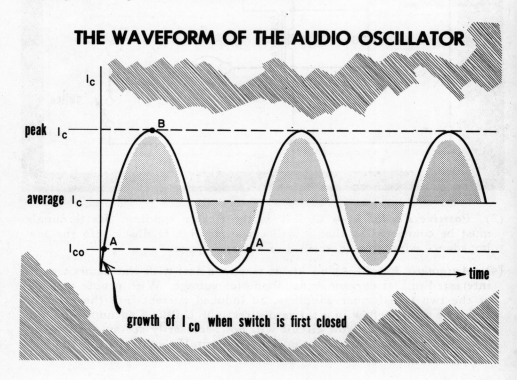

THE WAVEFORM OF THE AUDIO OSCILLATOR

growth of I_{co} when switch is first closed

Radio-Frequency Oscillators

Radio-frequency oscillators differ from oscillators of lower frequency in three major respects: first, the tuned circuit is made resonant to the high frequency desired; second, a transistor having a sufficiently high cutoff frequency must be used; and third, more care must be exercised in providing fixed bias to insure oscillation over a wide range. R-f oscillators in broadcast superheterodynes, for example, must be designed carefully in these respects if reliable performance over the entire broadcast band is to be obtained.

Many oscillators used in radios are built on the principle of the Hartley. A glance at the circuit will show that it is essentially the same as the

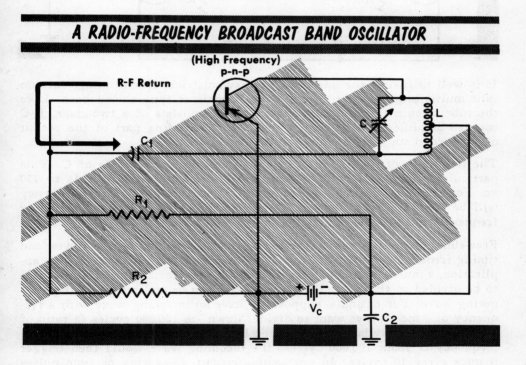

A RADIO-FREQUENCY BROADCAST BAND OSCILLATOR

vacuum-tube arrangement, involving a tuned circuit for the frequency desired, transistor amplification, feedback by the tapped-coil method and a biasing network. The L-C circuit is tunable from 995 kc to 2055 kc, the oscillator being designed for use in a radio having a 455 kc i-f.

Note that R1 and R2 are our old friends, the voltage-divider bias resistors which set the operating point of the oscillator firmly where desired. C1 provides a clear return for r-f collector currents back to the base. Similarly, C2 returns the collector r-f currents to the emitter with little impedance.

Transistor Multivibrator

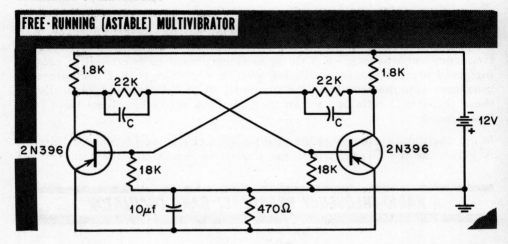

FREE-RUNNING (ASTABLE) MULTIVIBRATOR

It is well to review the operation of a symmetrical free-running, vacuum tube multivibrator before studying the transistor type shown here. Like the tube circuit, the transistor multivibrator consists of a two-stage, R-C coupled amplifier. Oscillation is sustained by feeding part of the output of the second stage back to the input of the first stage.

The frequency of oscillation is governed chiefly by the value of C. This particular circuit is capable of consistent operation from 1 cycle to 250 kc. For example, when both capacitors (C) are made 100 $\mu\mu$f, the frequency will be approximately 200 kc. If they are now raised to 1 μf each, the frequency will go down to approximately 40 cycles*.

Free-running multivibrators are useful for generating square waves and timing frequencies. They are often used as *frequency dividers*. In this application, a multivibrator is triggered by the waveform whose frequency is to be divided so that it oscillates only once for several pulses of the triggering wave. For example, suppose we have an incoming wave with a frequency of 1 mc that we want to divide "down" to 100,000 cycles (a ratio of 10 : 1). The multivibrator would then be adjusted so that its free running frequency is about 100,000 cycles; the incoming wave would then trigger it once every 10 cycles. In our sample circuit, triggering or sync pulses are fed to the multivibrator via a small capacitor going to the base of the transistor on the left. Output is taken from the collector of the transistor on the right.

---

* The equation that gives approximate frequency is

$$f \cong \frac{40}{10^{-6}(C + 100)} \quad \text{(C in } \mu\mu f)$$

Crystal Oscillator

Almost any vacuum tube crystal oscillator circuit can be made to operate when a transistor rather than a tube is used as the amplifier. This is particularly true of the Colpitts, Clapp and Miller oscillator circuits. Due to the differences between tube and transistor behavior, however, certain transistor crystal circuits that have no duplicates in vacuum-tube oscillators have become quite popular. One of the best of these is shown.

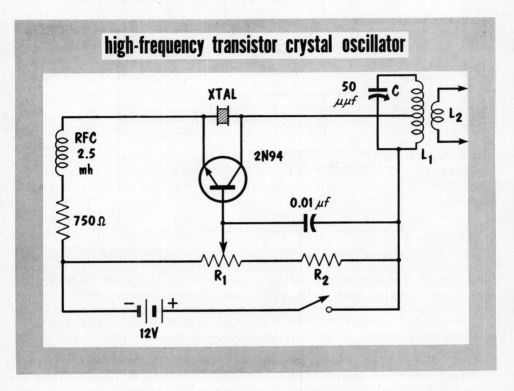

In this particular arrangement, oscillation has been obtained without difficulty to 5 mc using an r-f transistor such as the 2N94. With transistors designed for higher frequencies, there appears to be no question about extending the range to 10 or 20 mc using the same circuit. Note the center-tapped tank coil and the variable base resistor; both of these enter critically into the successful performance of the oscillator at very high frequencies.

The controlling crystal is connected directly from the collector to the emitter. If one were to draw an analogy to a vacuum tube circuit using this pattern, one would have to connect the crystal directly between the cathode and plate of the tube. There simply is no such crystal-controlled tube circuit; the reasons for this will become clear, shortly.

Crystal Oscillator (contd.)

COMPONENT FUNCTIONS IN CRYSTAL OSCILLATOR

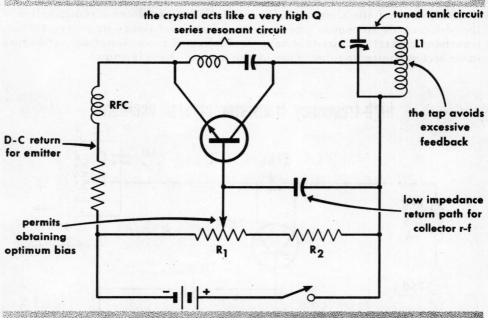

the crystal acts like a very high Q series resonant circuit

tuned tank circuit

C   L1

RFC

the tap avoids excessive feedback

D-C return for emitter

low impedance return path for collector r-f

permits obtaining optimum bias

R_1 R_2

The transistor in the crystal oscillator is connected in a common-base con-figuration. Looking back at the phase relationships between emitter input and collector output in common-base connections, we find that *there is no phase shift from input to output.* This is an important clue; it tells us why a transistor will work in this circuit and why a tube will not.

Assume that a small oscillatory current flows in the tank circuit $(C-L_1)$ when the switch is closed. The r-f voltage developed at the coil tap can then reach the emitter through the crystal. Most important is the fact that this feedback from the collector to the emitter is *positive* or regenera-tive feedback needed for sustained oscillation because there is no phase shift, as mentioned above. The fedback r-f is then re-amplified to supply the output energy as well as additional feedback for continued, undamped oscillation. In this circuit, the crystal acts as a series-resonant system having a very high Q, and establishes control of oscillation frequency by feeding back only one frequency for further amplification, i.e. *its own resonant frequency.*

The coil is "tapped down" to avoid feeding back an excessive amount of r-f current that might damage the crystal. Resistor R1 is made variable so that the base bias can be adjusted for optimum operation.

Transistor Switches

Modern computers, as well as a host of other control devices, often require that a relatively large load be switched on and off by a switch that cannot be made large enough to carry the necessary current. In other cases, switching is to be initiated by a current pulse from some previous circuit, the pulse amplitude being very small. Transistors are now enjoying widespread use as switching devices.

A basic switching circuit is shown in the illustration. With the small mechanical switch open, the load — which might be a 25-watt panel lamp or relay — is "off" because the only current flowing through it is I_{ceo} which

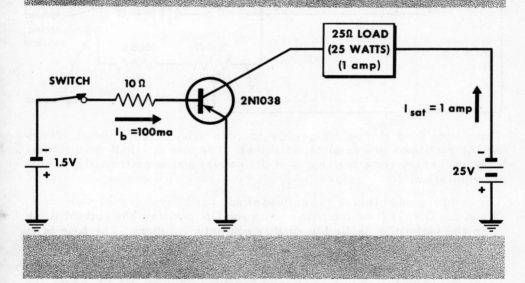

A TRANSISTOR SWITCH THAT CONTROLS 25 WATTS WITH ONLY 150 MILLIWATTS

SWITCH   10 Ω   2N1038   25Ω LOAD (25 WATTS) (1 amp)

$I_b = 100$ ma $I_{sat} = 1$ amp

1.5V   25V

is very small (a few microamperes). When the switch is closed, about 0.5 volt is applied as forward bias to the base from the 1.5-volt battery through the 10-ohm base current limiting resistor. The base current flowing under these conditions drops the emitter-collector resistance to a fraction of an ohm. Thus, the 25-volt battery is effectively connected directly across the load with very little voltage drop in the transistor.

A circuit such as this can handle a load power of 25-watts with an input power of only 0.15 watt, thus making it possible to trigger the load with a tiny mechanical switch or a small input pulse.

Time Delay Circuit

A TRANSISTOR TIME-DELAY CIRCUIT

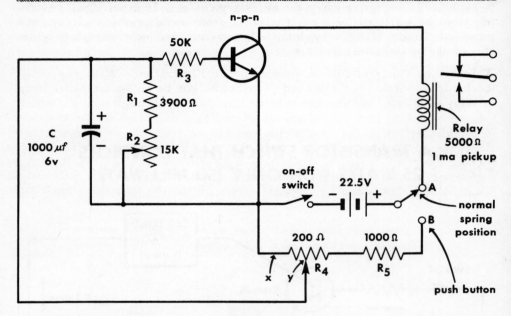

Transistors lend themselves readily to time delay circuit design. Many circuit variations are possible, of course. The one we shall discuss uses only one battery (or other source of d-c power) and is particularly flexible in application.

The control pushbutton is spring-loaded and normally occupies position **A**. When the ON-OFF switch is moved to the ON position, the current drain from the battery is negligibly small since only I_{ceo} flows. The base is at emitter potential through R3, R1, and R2.

To operate, the pushbutton is shifted to position B momentarily. This opens the collector circuit and connects voltage divider R4-R5 across the battery. Capacitor C charges to the potential between points X and Y, with the top end positive. When the pushbutton is released, collector current begins to flow, pulling in the 1-ma relay. For this condition, a large collector current is obtained because now the charged capacitor is slowly discharging through the base-emitter circuit in a forward direction, reducing the collector resistance. Since the capacitor is also discharging through R2, this potentiometer may be calibrated in seconds. At the end of the timing interval when C is discharged, the relay drops out. R4 is made variable so that the instrument may be kept in accurate calibration as the battery ages.

The Tetrode Transistor

One of the principal disadvantages of early junction transistors was their inability to perform well at high frequencies. As in a vacuum tube, capacitance between elements causes shunting of high-frequency signals both at the input and at the output. It has been shown, however, that capacitive shunting is *not* the most important factor in degrading the high-frequency response of a transistor. Of much greater significance is the effect known as *transit time dispersion*, sometimes given in terms of what is now called the "base spreading resistance". Let us see what transit time dispersion is and what its effects are.

In the common-emitter circuit, signal pulses are applied to the base electrode. Electrons moving into the base region from the emitter are therefore acted upon by an electric field that tends to increase or decrease the charge carrier intensity, depending upon the polarity at the instant in question. Unfortunately, various paths through the base are different in both length and quality. The net effect of this is that the transit time

Various paths through the the base region of a junction transistor have different transit times. This is called transit-time dispersion.

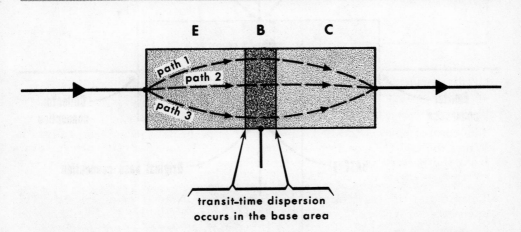

transit-time dispersion
occurs in the base area

Time required for carriers to pass through the base region along path 1 differs from path 2 time and both differ from the time for path 3.

through the base is not the same for all carriers that cross the emitter junction. This difference is quite small and is inconsequential at audio and low radio frequencies. At the higher frequencies, however, the difference becomes very important.

The Tetrode Transistor (contd.)

A given signal pulse at the transistor base is expected to cause the same change in carrier intensity at all points. Even if there is a delay in transit time, if that delay is uniform throughout, the waveform at the collector will not be changed materially from the input waveform. When the transit times are different in various points in the base, however, the output waveform at high frequencies tends to smear and seriously diminish in amplitude as a result of the dispersion of delay times.

One approach that has been used to overcome transit time dispersion is to make the transistor much smaller in physical size. By reducing the cross-sectional area of the base, the variation in delay times is not as severe be-

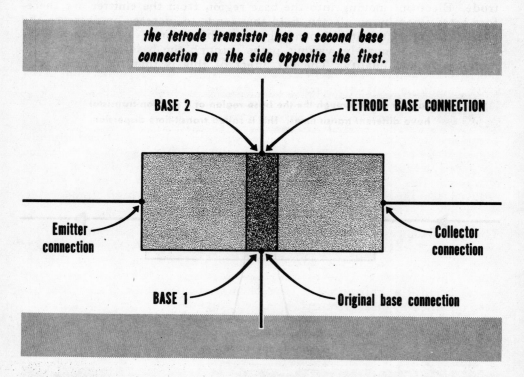

the tetrode transistor has a second base connection on the side opposite the first.

BASE 2 · TETRODE BASE CONNECTION

Emitter connection · Collector connection

BASE 1 · Original base connection

cause the number of different carrier paths is reduced in proportion. While this is of aid in solving the electrical problem, cost increases and reliability of performance suffers.

The tetrode transistor is the practical answer. As we shall show, the base paths may be greatly reduced in number by electrical means (rather than mechanical) while at the same time utilizing a transistor of reasonable size. In the tetrode, a second ohmic connection is made to the base on the side opposite the first base terminal.

The Tetrode Transistor (contd.)

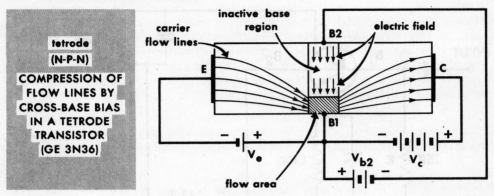

tetrode (N-P-N)

COMPRESSION OF FLOW LINES BY CROSS-BASE BIAS IN A TETRODE TRANSISTOR (GE 3N36)

carrier flow lines · inactive base region · electric field

A battery is now connected to the base terminals with a polarity such that the base 2 connection is more negative than the base 1 connection. This battery supplies *cross-base bias*. The electric field thus established "compresses" the flow lines of the charge carriers into one very small section of the base, leaving the upper portion inactive.

Cross-base bias in the tetrode transistor brings about a significant diminution of transit-time dispersion or base spreading resistance. Now the paths are greatly reduced in number so that transit times are equal for all active charge carriers. As a result, the smearing of the output signal is greatly reduced making increased power gain, even at very high frequencies, realizable.

The curve shows how cross-base bias affects the power gain of a tetrode transistor. Note that the gain is very small when the cross-base bias is near zero. In this situation, the transistor behaves very much like the conventional three-terminal type as the bias is raised. The power gain also increases until a limiting point is reached at about 2 volts. This curve has been obtained by measurements taken at 60 mc.

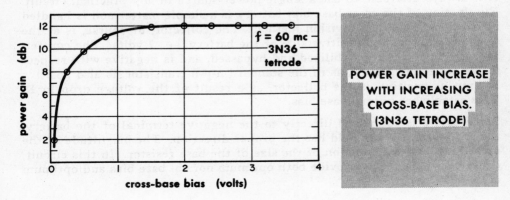

f = 60 mc 3N36 tetrode

POWER GAIN INCREASE WITH INCREASING CROSS-BASE BIAS. (3N36 TETRODE)

Biasing the Tetrode Transistor

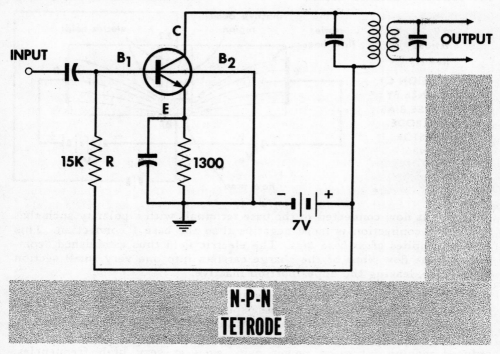

For a typical transistor of the tetrode variety, the following voltages and currents are recommended:

V_{ce} (collector-emitter voltage) = 5 volts
V_{b1b2} (base 1-base 2 voltage) = 2 volts
I_{b1b2} (base 1-base 2 current) = 0.5 ma
I_e (emitter current) = 1.5 ma

It is always desirable to use a single power source in any practical circuit. Elimination of the cross-base bias battery is a simple matter and is handled by a standard voltage division method. The collector, of course, is at the same potential as the positive side of the battery, i.e., 7 volts. The emitter, as usual, is resistance-stabilized and bypassed, and is negative with respect to the collector. Base 1, as in the standard n-p-n transistor, is also positive but not as positive as the collector, as a result of the voltage drop in R. This establishes normal base bias.

Since base 2 is connected directly to the negative terminal of the battery, the cross-base electric field has the proper direction. The magnitude of the cross-base bias is a function of the size of the base resistor. In this circuit, it has been selected to provide both optimum normal base bias and optimum cross-base bias.

An Interesting Application of the Tetrode Transistor

We have seen how the power gain of a tetrode transistor can be varied from 1 or 2 db all the way up to 12 db or more at a high frequency such as 60 mc, merely by changing the cross-base bias.

This immediately suggests that cross-base bias might be used for automatic gain control (agc or avc) in high-frequency r-f or i-f stages in either radios or TV sets. Transistorized receivers have been prone to become detuned with standard agc circuits. The circuit in the diagram offers an extremely stable and reliable method of signal-control of gain without noticeable detuning.

The system depends upon a negative d-c detector signal as in any avc circuit, this signal being obtained from the detector of the receiver. With a weak-signal at the antenna, the negative control voltage applied to the base of TR2 is small, TR2 is virtually cut off and the top end of R5 is almost at ground potential. This establishes sufficient cross-bias to bring the tetrode (TR1) into the high-gain region. As the signal intensity increases, the control voltage becomes more negative, TR2 conducts more heavily and a voltage drop having the polarity shown on R5, is developed. This reduces the cross-base bias, decreasing the gain of the tetrode.

USING CROSS-BASE BIAS IN AN AUTOMATIC-GAIN-CONTROL CIRCUIT

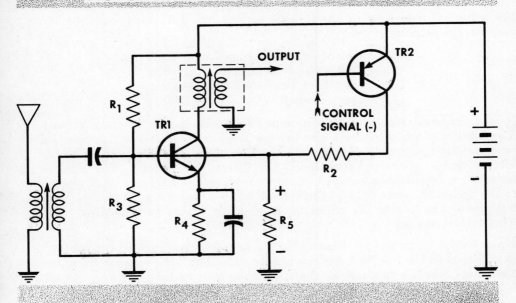

QUESTIONS AND PROBLEMS

1. Name the parts of any oscillator, giving the function of each one.

2. Draw a diagram of an L-C transistor oscillator. Identify the frequency-determining network, the amplifier, the place where positive feedback occurs, and the bias components.

3. Using a diagram to help you, explain fully the operation of a transistor audio oscillator.

4. What precautions must be taken to insure reliable operation when assembling a transistor r-f oscillator?

5. Draw a diagram of a transistor multivibrator of the astable type and explain the sequence of steps that take place in a single cycle.

6. How does a transistor crystal oscillator differ from a tube type of crystal oscillator? Explain how this difference comes about.

7. What is meant by a transistor switch?

8. Draw a transistor switch circuit and explain how it works.

9. Describe what is done in the fabrication and use of a tetrode transistor that makes it suitable for use at high frequencies.

10. Explain the operation of a tetrode agc circuit.

INDEX

Absolute maximum current, 50
Absolute maximum rating, 48, 49
Acceptor, 22
Acceptor impurity, 18
A-C generator, 24
A-C signal amplifiers, 99
Added impurity effect, 15
Adding input and output, 115
Alpha cutoff frequency, 52
Alpha, meaning of, 40
Ambient temperatures, 26, 85
Amplification, 128, 129
Amplifier, 1
Analyzing the coupling, 114
Antimony, 14
Application of the tetrode transistor, 141
Arsenic, 14
Atomic number, 7
Atomic particles, 5
Atomic weight, 7
Audio amplifier: 95
    using complementary coupling, 117
Audio oscillator operation, 130
Audio preamplifier, 108
Automatic bias, 128, 129
Automatic gain control, 141

Bandpass amplifiers, 99
Bardeen, J., 1
Base bias, 89
Base-collector junction, 38
Base current: 37, 68, 70
    gain, 75
Base 1 — base 2:
    current, 140
    voltage, 140
Base-to-collector circuit, 38, 41
Battery polarities, 68
Base spreading resistance, 137
Beta, meaning, 69, 70, 71
Bias resistor, 66, 80
Bias, transistor, 34
Bias voltage, 125
Biasing:
    n-p-n transistors, 38
    p-n-p transistors, 38
    the tetrode transistor, 140
Black graphite, 11
Blocking capacitor, 108
Brattain, W. H., 1
Breakdown point, 25
Bucket brigade, 15
Bucking voltage, 85

Capacitive shunting, 137
Carbon, 11
Carrier diffusion, 56

Cathode, 39
Cathode bias, 66
Cat whisker, 31
Charge carriers, 18, 39
Characteristic curve, 103
Chemical properties, 10
Circuit configuration, 57, 58, 59, 60
Clapp oscillator circuit, 133
Collector: 36, 39
    characteristics, 76
    characteristic curves, 72
    current: 37, 40, 50, 73, 74
        with emitter circuit open, 37
    cutoff current, 56
    curves, 72, 73, 74, 75
    -emitter voltage, 140
    load power, 44
    -to-base resistance, 84
    supply voltage, 67
    voltage, 73, 77
Colpitt oscillator circuit, 133
Common base: 57
    amplifier, 63
    configuration, 45, 79, 93
Common-collector: 60
    amplifier, 88
    circuit, 87, 88
    n-p-n transistor, 87
    p-n-p transistor, 87
    single source circuit, 89
Common compound, 9
Common emitter: 58, 59
    characteristics, 72
    circuit, 58, 68, 78
    configuration, 69, 93
    current gain, 70
    to common emitter, 110, 111, 112
Common ground, 67
Common reference, 66
Comparing a transistor to a triode, 39
Comparison of frequency responses
    of configurations, 93
Complementary condition, 113
Complementary nature of transistors, 113
Complementary symmetry, 123
Complex atoms, 6, 7
Computers, 97
Cost and space economy, 106
Conduction, 47
Conductors, 8
Contact resistance, 97
Control elements, 57
Convection, 47
Copper, 11
Coupling:
    impedances, 100
    methods reviewed, 99, 100, 101
    system, function of, 102, 103, 104, 105

INDEX

Covalent bond, 12, 15
Cross-base bias: 139
  battery, 140
Crystalline, 11
Crystal oscillator, 133, 134
Current:
  amplification, 83
  flow through a P-N junction, 22, 23
  gain, 52
  gain squared, 79
  -operated amplifiers, 104
Cutoff:
  current, 83
  frequency, 125

D-C amplifier, 94
D-C bias voltages, 107
D-C continuity, 81
De Forest, Lee, 1
Degeneration, 117
Degenerative effect, 85
Demodulation, 96
Diamond, 11
Diamond crystal, 12
Diffused-alloy process, 19
Direct coupling, 110, 111, 112
Distortion, 106
Donor impurity, 18, 22
Double-tuned transformer, 101
Driving power, 121, 122

Effect of a thin central section, 35
Effect of the second battery, 36
Efficiency, 121, 122
Electrical characteristics, 48
Electrical conductors, 8
Electrical insulators, 9
Electrons: 1, 5, 25
  and holes, 17
Electron configuration, 10
Elements, 4
Emitter: 36, 39
  -base impedance, 67
  -base junction, 37, 89
  battery, 37
  bias voltage, 40
  current, 50, 68, 140
  stabilizing: 121
    resistor, 96, 108
  -to-base circuit, 38, 41
  -to-collector current gain, 40

Feedback network, 83
Feedback resistor, 85
Filament power, 24
Forming a junction, 19, 20, 21

Forward bias, 34, 41, 66, 68, 80, 85, 106
Free electrons, 8
Frequency:
  determining network, 128, 129
  dividers, 132
  of oscillation, 132
  response, 62, 63, 92, 107, 121, 122
Fused rectifers, 33

Gain, 106
Gaseous diffusion method, 21
General considerations for amplifiers, 62
Germanium: 11, 13, 26
  diode, 23, 24
  -indium alloy, 19
  junction rectifier, 24
  pellet, 26
  rectifier: 28, 33
    curve, 25
  transistor, 46
Grid, 39
Grounded-base connection, 45
Grounded cathode, 58
Grounded-emitter circuit, 58
Grounded-grid amplifier, 57

Hartley oscillator, 128
Heat, 2
Heater current, 2
Heat sink, 21, 47, 119
Heat transfer method, 47
Helium, 6
High-current silicon rectifier, 27
High-current bridge rectifier systems, 30
High fidelity:
  audio amplifiers, 54
  systems, 28
High frequency: 48
  coupling in practice, 125, 126
High-frequency response, 95
High-impedance primary, 67
High-purity germanium, 35
High-resistance direction, 41
Hole concept, 17
Holes, 16, 25
How the power supply works, 29
Hydrogen, 5

Impedance:
  average for a 2N78, 90
  coupling: 104
    network, 100
  in the common emitter, 78
  matching, 104
  to common-collector circuit, 90
  transformer, 92

INDEX

Indium: 16
  bead, 26
Induced voltage, 100
Industrial power systems, 21
Inert elements, 9
Ingot, 20
Input:
  current, 70
  impedance: 62, 63, 78
    matching, 78
  resistance, 42
  transformer, 67
Iodine, 11

Junction: 23
  capacitance, 32
  diode, 31
  rectification, 24
  temperature, 51

Lattice, 12
Line voltage, 29
Lithium, 7
Loading resistors, 101
Low-current germanium rectifiers, 27
Low-frequency response, 95
Low-input impedance, 92

Majority carriers, 25
Manganese, 11
Matter, 3
Maximum collector dissipation, 51
Maximum inverse peak voltage rating, 27
Maximum peak inverse voltage, 29
Maximum rating, 29
Mechanical shock, 2
Metals, 8
Microwave mixer, 32
Miller oscillator circuit, 133
Minority carriers, 25
Modulator transformer, 104
Molecular building blocks, 4
Molecules, 4
Molten germanium, 14
Multistage amplifiers, 65

Neutrons, 5, 6
Nickel-steel spring, 26
Noise figure, 55
Nonconductor, 22
Non-crystalline form, 12
N-P-N:
  assemblies, 20
  transistor, 33
Nucleus, 6

N-type germanium, 18

Objectives in selection of coupling system, 106
Operating point, 121
Orbit, 7
Oscillatory current, 134
Outer shell, 8
Output:
  capacitance, 55
  impedance: 53, 63, 78
    matching, 78
  power, 44
  resistance, 42

Parallel resonant circuit, 126
Pentavalent atom, 14
Pentavalent impurity, 18
Phase:
  inversion, 65, 66
  relations:
    in a common-base amplifier, 65
    in the common-emitter circuit, 86
Physical characteristics of transistors, 46, 47
Physical laws, 3
Physical structure, 119
Photocurrent, 94
Plate, 39
P-N-P transistors, 19, 33
Point-contact diode, 31
Poor conductor, 12
Positive charge, 16
Positive electrical charge, 5
Positive feedback, 128, 129
Power: 2
  amplifier, 30, 104
  gain, 79, 91, 121, 122
  handling ability, 46
  output, 121, 122
  per unit-time, 118
  source, 96
  transistors: 30
    complementary symmetry, 123, 124
    push-pull amplifiers, 122
    typical characteristics, 120
    what makes them different, 118
Practical biasing, 84, 85
Practical common-base amplifier, 67
Preamplifiers, 28
Protons, 5
P-type germanium, 18, 20
P-type wafers, 20

Questions and problems, 32, 61, 98, 127, 142

Radiation, 47
Radiation surface, 119

145

INDEX

Random molecular movement, 56
Rate grown transistor junction, 20
Rate growth process, 20
Rating charts, 48
R-C coupled transistor amplifier, 108
Recombination process, 40
Rectifiers, 26
Relay, 94
Resistance-capacitance coupling, 100
Resistance gain, 42
Resonant frequency, 134
Reverse bias, 34, 38, 41, 68, 80, 87
Ribbed structure, 119
Ring, 7
Ripple voltage, 30

Sample circuits, 95, 96, 97
Sandwich, 20
Selenium barrier-layer, 116
Semiconductor rectifier: 33
    ratings, 27
Semiconductors, 10, 11, 12, 13
Shell, 7
Shockley, W., 1
Short-distance forces, 8
Signal:
    characteristics, 48, 52, 53, 54, 55
    control, 141
    generator, 42
Silicon: 13, 26
    solar cells, 116
Silver, 11
Simple power supply for transistorized devices, 28
Single-battery common-base circuit, 66
Single-battery common-emitter circuit, 80, 81, 82
Single-ended amplifier, 121
Slug-tuned transformer, 125
Special characteristics, 48
Stabilizing emitter resistor, 107
Static biasing, 89
Storage temperature, 51
Sustained oscillation, 128
Switching operation, 48
Symbolization, 45

Temperature: 46
    effect, 85
    factors, 51
Tetrahedron, 12
Tetravalent atom, 14
Tetrode transistor, 137, 139

Thermal agitation, 85
Time-delay circuit, 136
Transformer-coupled transistor amplifier, 107
Transformer coupling, 100
Transformer, secondary, 107
Transistor:
    action, 34
    arrangement, 58
    audio oscillator, 129
    configuration, 57, 83
    materials, 11
    multivibrator, 132
    parameters, 84
    resistances, 41
    switch, 97, 135
    wafers, 20
Transit time: 139
    dispersion, 137
Triode tube, 1
Trivalent atoms, 16
Trivalent impurity, 17, 18
Tuned amplifiers, 99
Two-battery positives, 81

Universe of matter, 4
Untuned amplifiers, 99
Untuned coupling, 100
Using the common collector amplifier, 92

Vacuum tubes: 2
    circuit, 57
    phase relations, 64
Valence electrons, 8, 11
Vest-pocket receiver, 96
Video amplifiers, 65
Video second detectors, 32
Voltage: 22
    amplification, 30
    and power gain in the common emitter, 79
    drop, 83
    gain: 43, 79, 91
        of a transistor, 43
    polarities, 59

Warm-up time, 24
Why more thought must be given to biasing, 83

X rays, 8

Zero-signal collector, 125
Zero-signal current, 83